A HOUSE
FOR THE F

TERENCE MCLAUGHLIN

Illustrations of
the Macclesfield house
by Brian Watson

TVTimes
FAMILY BOOKS

INDEPENDENT TELEVISION BOOKS LTD, LONDON

INDEPENDENT TELEVISION BOOKS LTD,
247 Tottenham Court Road,
London W1P 0AU

ISBN 0 900 72755 1

Printed in Great Britain by
St. Stephen's (Bristol) Press Ltd
Poplar Road, Warmley, Bristol

Also available in this series:

BARRY BUCKNELL ABOUT THE HOUSE
Barry Bucknell

BEATING THE COST OF COOKING
Mary Berry

CARAVANNING
Barry Williams

DEAR KATIE
Katie Boyle

HOUSE PLANTS MADE EASY
Jean Taylor

KNOW YOUR CAR
John Dyson

KNOW YOUR RIGHTS
Dr Michael Winstanley
and Ruth Dunkley

MAKING THINGS FIT
Ann Ladbury

POWER TOOLS AT HOME
Harold King

'WOMEN ONLY' COOKBOOK
Compiled by Barbara Twigg

CONTENTS

1 THE SELF-SUFFICIENT HOUSE—
WHAT IS IT ALL ABOUT? 4
2 POWER FROM THE SUN 7
3 WIND POWER 25
4 WATER POWER 34
5 THE HEAT PUMP 36
6 METHANE GENERATORS 39
7 RECYCLING WASTE 41
8 RECYCLING WATER 42
9 WHAT CAN YOU DO NOW? 44
10 PLANNING—WHO NEEDS TO KNOW? 46
11 INSULATION 50
12 MAKING USE OF ALTERNATIVE TECHNOLOGY
by Brian Trueman 62
FURTHER INFORMATION 77
ACKNOWLEDGEMENTS 78
INDEX 79

Chapter 1
The Self-sufficient House—What is it all about?

Most people think of houses as boxes to live in—some large, some small, some crowded together and others standing in splendid isolation, but boxes all the same. You put the boxes down by the side of a road, fill them with people, and they remain, static shells, unconnected with the life that goes on in and round them.

In reality a house is not like a box at all, but far more like a complex living organism. It takes in a whole diet of services—electricity, gas, water, communication in the forms of post, telephone, radio and television, and all the other supplies for the business of living—and discharges its wastes—sewage, rubbish and dirty water. To keep this organism alive, the house is connected to the outside world by an intricate network of lifelines—electricity cables, gas pipes, water mains, drains, wires, channels, and all the access routes necessary for the delivery of goods and the removal by the dustmen of the inevitable waste products.

If any one of the major lifelines is cut off for very long, the house organism begins to die, as all of us know from experience of power cuts, floods or even such a common event as a blocked drain. And it is in

Beginnings—Geoff and Lynn Grant with 3 year old Helen at their new home

The house nearing completion

these periods of shortage that we realize just how wasteful and greedy the average house is as an organism. We transport water from the purification works, sometimes over great distances, make it slightly dirty—at the rate of 159 litres a day for every one of us—and then send it back to the sewage works to be cleaned up again. We pipe our sewage away often as far as the sea, and then buy huge quantities of artificial fertilizers that themselves use up precious energy in their manufacture. We take in oil, coal and electricity, to use for heating in houses so badly insulated that three-quarters of the heat escapes into the open air.

When fuel was plentiful and cheap, this did not seem to matter, but recent events have taught us how short-sighted that policy was and how near we are to the limits of our resources. Not only are we in Britain spending over £10 million every day on imported fuel and thus crippling our economy, but on a worldwide basis the human race every year is burning up deposits of fossil fuel such as coal and oil that took about a million years to form. We squander our other natural resources in the same way. In Britain, for instance, despite the wet climate, every summer sees us nearer to the crisis point where we would use more water than our reservoirs can hold, and to satisfy our taste for wasting water valuable farmland has to be flooded to make more capacity.

A self-sufficient (or autonomous, or autarkic) house changes all that. The sun provides all the heat needed for comfort and cooking, and the house is insulated to the point when only a tiny fraction of the heat is wasted. As electricity is not needed for the heavy demands of heating, a gas- or wind-powered generator can cope with the electrical needs of lighting, household machines and tools, radio and TV, etc. If there is any heat or power to spare this can be taken to a greenhouse or a patch of garden warmed by soil-heating cables, so that early vegetables get a chance and the food supply can be kept up all the year round. Sewage and kitchen wastes are processed to make compost for the garden, or fermented to produce methane gas that can be used as fuel for supplementary heating plant or generator, or even used to run a car. All the water used for washing, bathing, and so on is recirculated through purifiers and filters and used again; rain collected from the roof makes up losses. (If the idea of using your own bathwater, even purified, for drinking is repellent to you, remember that if you live in any of the heavily populated parts of the country you are almost certainly getting through your taps someone else's bathwater which has been discharged to the rivers and then repurified.)

Such a house is not just a science fantasy. All over the world people are designing and building self-sufficient houses and getting to grips with the practical problems of running them, and the hardware that is necessary to keep them going. Geoff and Lynn Grant and their two children are one of the families finding out just how to live a normal life without draining away precious resources of energy, water, and other irreplaceable materials. Their house, designed by architect Donald Wilson with the backing of Granada Television, has been constructed from the shell of an old coach-house near Macclesfield. With the help of a team from Granada, a variety of volunteers, and Don Wilson not only designing but working alongside them, the Grants are trying out all the latest ideas in autonomous house management—living in a house that is not only a home but an experiment in the new art of living that we shall all have to cultivate if civilization is to survive.

While we cannot all have the opportunity to design the house for the future from an empty shell, or to build from the ground up, as some groups have been able to do, there are ideas from these experimental houses that can be adapted to the most conventional home to enable people to live a fuller life while at the same time saving their own money and the country's resources.

Chapter 2
Power from the Sun

The sun is the ultimate source of almost all our energy, and the mainstay of life on earth. Its light and heat are essential for the growth and continued existence of every plant and animal; even when it is obscured or hidden as in the depths of winter or at night, the fuels we burn as substitutes for its warmth are only themselves forms of stored-up solar energy. Wood and peat, for example, are the products of tens or hundreds of years of sunshine on trees and grasses; coal and oil, the fossil fuels, are the buried memories of millions of years during which the sun encouraged the growth of prehistoric plants and animals.

The total amount of energy radiated by the sun is, of course, millions of times greater than any quantity we could dream of using, but most of it is directed towards outer space, not to the planets, which are little more than dots on the horizon as seen from the sun. Even so, the amount of energy that actually comes in the earth's direction is impressive—at the outer layers of the atmosphere solar power is received at about 1,300 watts per square metre. Most of this is absorbed or

Solar panels under test at Pilkington Brothers Research and Development Centre

The Macclesfield solar roof collector completed and glazed

reflected by the layers of atmosphere, clouds, etc., which is perhaps just as well, for if the full heat of the sun reached the earth's surface the seas would boil and every living thing be reduced to a crisp.

Even in the cloudy latitudes of Britain, however, the average amount of solar energy that penetrates the atmosphere and arrives at ground level is more than 100 watts per square metre, which means that the total solar energy falling on the British Isles is around 20 million megawatts, or 400 times the present output of all our power stations put together. In warmer countries, the potential of solar energy is even greater: it has been calculated that the sun radiates more power on Australia *every year* than the amount of energy we could extract from all the known reserves of coal and oil in the world.

These figures underline the essential paradox of our so-called energy crisis: there is no energy shortage, in real terms. It is just that we insist on generating our power from a very restricted set of sources—fossil fuels and nuclear materials—that are fast running out. If we could collect the sun's energy with only partial efficiency on any large scale, we could have more power than our most extravagant current use, without pollution, radioactive dangers or any imminent risk of exhausting the supply.

Collecting solar energy

The best way to collect the sun's radiation would obviously be to have a panel, mirror or lens which moved around to face the sun at all times, but this would be a difficult engineering problem for a large collector. In practice, most systems manage quite well with a fixed collector, facing south and angled to face roughly halfway between the sun's highest elevation and its lowest. The collector should have a matt black surface to absorb the maximum amount of heat (black plastics are often used, as they can also protect the structure underneath from the weather, but there are special black paints available, usually based on copper oxide, which have very high absorbency for solar radiation and can be used over any surface). Such a system is called a 'flat-plate collector'.

Obviously it is necessary to have some means of carrying the heat from the collector to the house. Most systems depend on circulating water in channels or tubes through the collector, and passing it through pipes to the interior of the house, where it can be collected in a storage tank or circulated through radiators as desired. Some designers favour a 'closed' system with the water in tubes all the way (as in a conventional central-heating system, but working in reverse—taking in heat instead of putting it out), while others use open channels with water simply trickling over the hot surface of the collector. Whichever system is used, it is necessary to cover the absorbent surface with a layer of glass or, ultra-violet resistant, transparent plastic. This not only protects the collector from weather (and, in the open systems, prevents the hot water from evaporating and thus wasting heat), but acts as a kind of 'greenhouse' to concentrate the heat.

This 'greenhouse effect' depends on the fact that the sun's heat comes to earth in the form of fairly short-wave infra-red radiation, that can easily pass through glass or plastic. Under the glass, objects get hot and try to radiate their new-found energy back again towards the sky. Fortunately, however, as they are infinitely less hot than the sun, their radiation comes off as very long-wave infra-red, and this cannot pass through glass very easily. So the heat from the sun is trapped by the layer of glass. Of course some heat escapes by leakage through joins in the glass panes or simply because the glass gets warm all through and actually conducts heat away, but most of it is retained. This is the principle behind double-glazing, greenhouses, garden cloches, and similar ways of trapping the sun's heat under glass or plastic sheets.

'Open' systems

Donald Wilson's collector design is typical of the 'open' system, and is adapted from Robert Vale's design for the Biotechnic Research and Development (BRAD) commune's home in mid-Wales. The whole southern roof of the house is made of 0.7mm corrugated aluminium, with the channels formed by the corrugations running down the slope of the roof. These sheets are supported on a series of short steel purlins that run from end to end of the house.

Steel was chosen instead of wood so as to withstand the very great differences in temperature expected between the top and bottom edges of the purlin section. This is because the top edge lies just behind the hot collector while the bottom edge is protected from the heat by insulation. Short lengths were used, instead of the more usual continuous purlins, to localise movement through expansion and contraction of the aluminium sheets in extremes of temperature—between a hot sunny day, when the roof is absorbing large quantities of heat, and a frosty night. The use of two metals together posed other problems. Corrosion can occur between dissimilar metals in the presence of moisture, and the steel purlins might also have conducted away some of the heat from the collector. To overcome both problems, 10mm thick packing strips were cut from asbestos board and placed between the aluminium and the steel. Asbestos cement was not used, as it might have become brittle and disintegrated, thus making the roof fixings loose. Silicone rubber

General view of roof from west

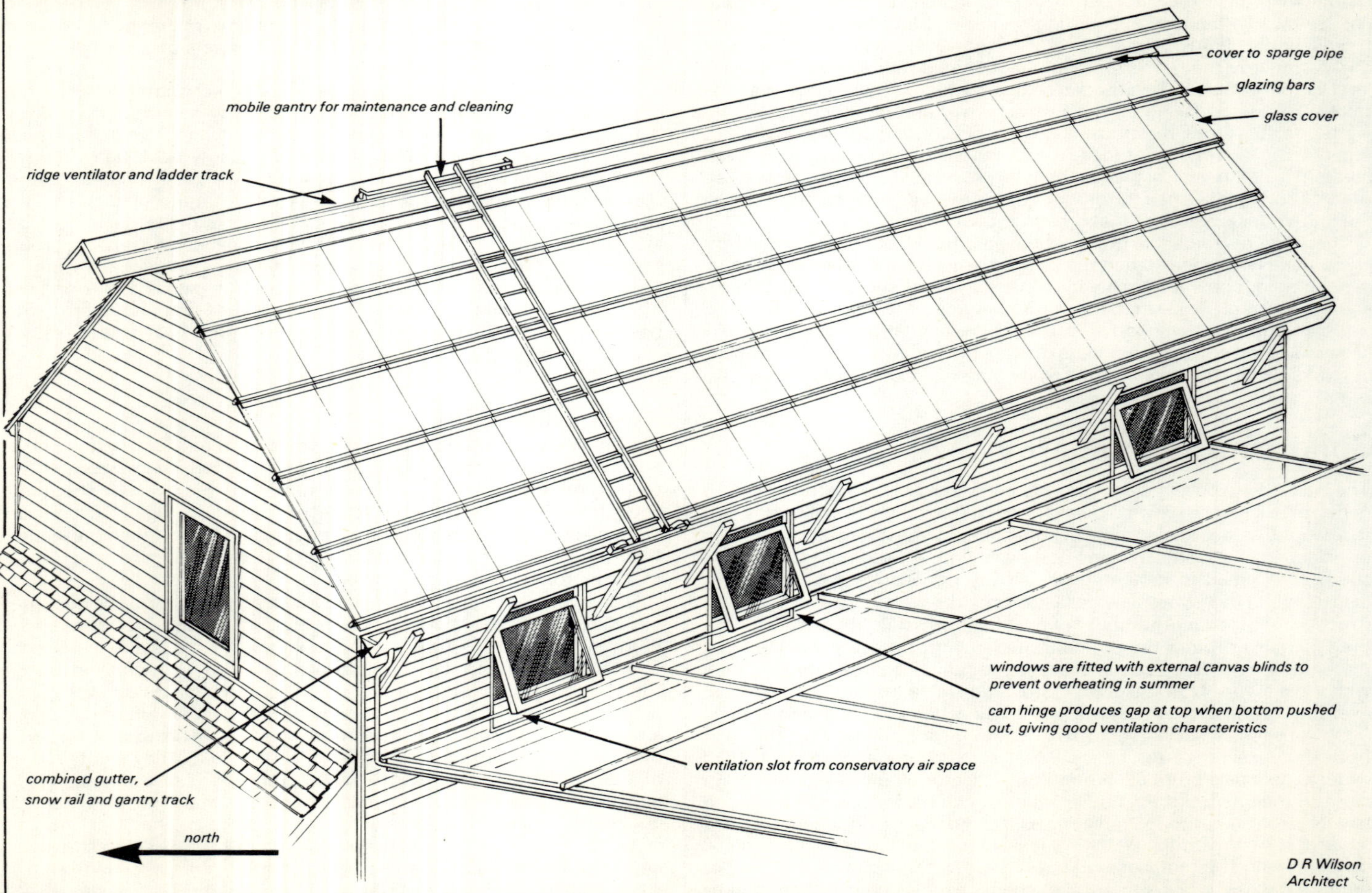

Solar roof construction

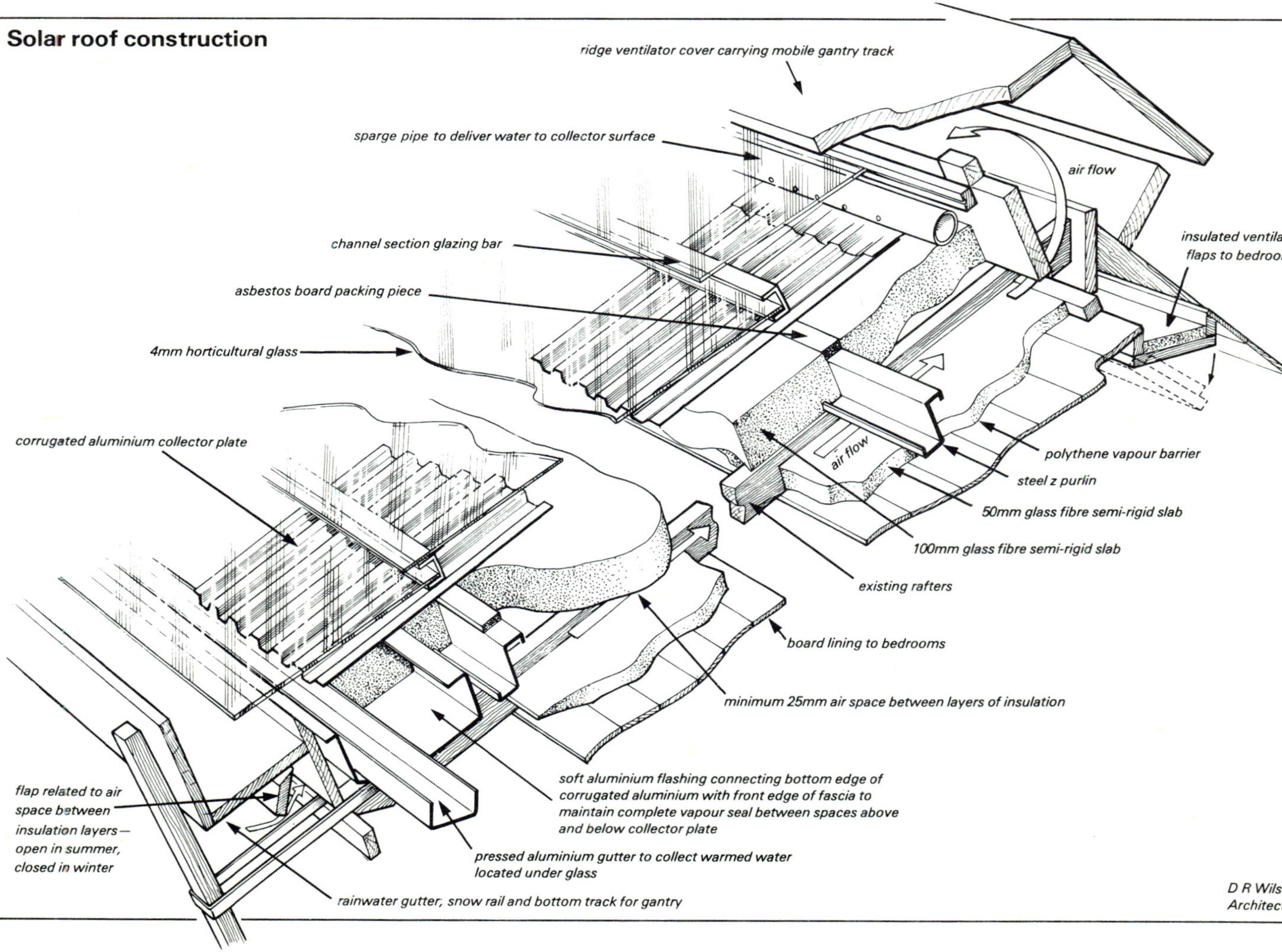

D R Wilson
Architect

was used to seal all the joints in the aluminium to prevent any of the moisture or vapour above the collector penetrating the insulation below. The collector then serves effectively as a vapour barrier, and prevents excessive moisture in the underlying roof structure.

The aluminium (Korrugal TRP20) is painted on the outer side with a matt black acrylic finish that serves both to absorb the solar radiation and to protect the metal from corrosion by the water which flows down the channels from a supply pipe running the length of the roof peak. As this water trickles down, it is warmed by the hot metal, and is collected at gutter level by a channel which carries it to storage tanks. A pump in the system keeps the water circulating at such a rate that it picks up the heat from the roof about as fast as it is absorbed.

To maintain the greenhouse effect, the whole roof is covered with 4-millimetre horticultural glass sheets, which not only trap the solar heat, but stop the trickling water from evaporating to the outer air, which would mean not only loss of water but wastage of heat. Spacer bars mounted on the upper surface of the channelled aluminium sheets hold the panes of glass in such a way that they are separated from the wet surface of the aluminium and also from one another. This means that they can overlap, for efficient shedding of rain or snow from the roof, without the risk of capillary attraction between two adjacent glass surfaces. This makes the glass roof more weatherproof and stops the growth of algae between the panes. This growth is often seen as a dirty green stain in the overlapping areas of panes of glass in green-

Biotechnic Research and Development (BRAD) headquarters in Mid-Wales, showing the 60m² solar roof/collector

houses. In the context of a solar roof it would not only be unsightly, but would cut down the transmission of the sun's heat. Silicone rubber seals are used on all the joints between glass and the aluminium mounting bars.

Closed systems

Closed systems for solar collection are more suitable for a house with an existing roof, as they can be installed over the tiles or slates without any major reconstruction. Often the panels of the system are hung from the ridge of the roof to give stability without the need to remove tiles; clamps at the bottom of the panel act as anchors. The collectors of closed systems are usually panels of metal or plastic fitted with pipes or channels to carry the circulating water that picks up the absorbed heat. For example, one popular pattern consists of an aluminium frame supporting an oil-tempered hardboard base, heavily insulated from the building and the surrounding air by rock wool. This base supports a blackened metal panel which has fixed to it a zig-zag array of pipes through which water can circulate to pick up the heat absorbed by the panel. The whole array is covered by a sheet of 4-millimetre glass to trap the solar heat.

Other closed system panels are made as a labyrinth of channels through which water can circulate inside a collecting 'slab'. One such panel consists of two layers of black polyvinyl chloride (PVC) welded into a panel with a maze of water channels in between the layers. The PVC is protected against rotting and has a special ingredient to make it resistant to the effects of ultra-violet light from the sun, which tends to make plastics brittle. In this way a panel 1.93-metres by 0.8 metres can be made with an actual length of waterway up to 58 metres, ensuring very close contact between the water and the heated plastic. The manufacturers claim that the absorbency of this panel is such that it is unnecessary to have a further layer of glass or transparent plastic over the top, although this can be added if desired.

Types of solar heater—advantages and disadvantages

So many designs have been put forward for solar heaters that it may be as well to examine the pros and cons of the most commonly-used types.

The first decision to make is whether to use an 'open' system, built into the house, or a closed system, with separate panels. The main advantage of all the open systems is that they are cheaper, especially if they are constructed as part of the new building. After all, every building must have a roof and it need cost very little more to make this a heat-absorbing roof. On the other hand, they have as a class two fundamental disadvantages. First, it is difficult to arrange for them to be at the ideal angle for maximum heat collection, because most modern roof designs have a pitch of 40° or less to the horizontal. If the roof is built at the right angle for efficient winter collection, around 65-70°, it becomes very steep and requires quite a lot more material (rafters, roof covering, etc.), so is expensive. Second, they tend to become uncomfortably hot in the summer, and this heat is transmitted into the house. Some designs have white or reflecting covers which can be drawn over the collectors if the heat gets excessive, but this means having a fairly complex mechanism to move the covers because you do not want to have to clamber around on the roof fixing covers every time the weather changes.

Among the various materials proposed for fixed collectors, by far the simplest is concrete, which is used in several designs for swimming pool heaters. The water from the pool is pumped up over a large concrete ramp facing the sun and is warmed as it flows over the hot masonry. You will probably have noticed the same effect on sunny days at the seaside: if you bathe in the shallow water as the tide comes in, it is always pleasantly warm from flowing over the hot sand.

Concrete in thick sections, however, is very expensive for a roof, as is copper, and the best material economically seems to be aluminium. It is reasonably light while being strong structurally and having good resistance to the corrosion effects of the weather and flowing water. Galvanized steel (corrugated iron) is as strong, but heavier and not so easy to proof against corrosion. Most corrugated plastics are neither strong enough nor sufficiently resistant to ultra-violet light from the sun to be used successfully in any roof designed to last, although several experimental systems have been made with such materials.

Panels for closed systems are more expensive than the open roof, but can easily be set up at the right angle for optimum collection of solar power, and, if they are slightly separated from the actual roof of the house, do not overheat the building in summer. They can also be fitted easily into an existing hot water system.

Of the materials available for panels, copper is probably the best, though the most expensive. It can easily be treated to give it a black highly-absorbent surface (copper oxide, which is used in most of the special paints to make other surfaces absorbent to heat), it can be plumbed into a water system using simple soldered or compression joints, and its heat-transfer characteristics are very good, so that all the heat absorbed is passed rapidly to the circulating water.

Aluminium is almost as good as copper and cheaper, but it is more difficult to make good plumbing joints in the material and needs careful treatment to avoid corrosion. Aluminium collectors should not be used in conjunction with copper pipework.

Plastics such as black polythene or PVC are used widely and have one great advantage over the metals—they are slightly elastic and are therefore not normally affected if the water freezes. However, their heat transfer is not nearly as good as that of metals, and they must be very carefully treated to prevent them from becoming brittle when exposed to the ultra-violet rays of the sun. Plastics with an 'ultra-violet inhibitor' are used for panels, but the life of these materials is still somewhat in question.

Nearly all solar panels and roofs require a transparent covering to achieve the 'greenhouse effect', and the choice here is effectively between glass and plastic. Glass is heavy and can easily get broken, but on the other hand most of the transparent plastics are badly affected by ultra-violet light, becoming brittle and opaque. A material called Mylar is better than most other plastics in this respect, but the useful life even of Mylar when exposed to strong sunlight is a lot less than that of glass, probably about 10 years. Most designers of solar collectors seem to be playing safe and specify glass covers for their equipment, despite the risk of breakage.

Circulating the water

Whatever the system used, the circulating water can either be taken directly indoors by suitable pipework, or a series of the panels can be connected together by short lengths of tube so that the water circulates through all of them in turn, gradually getting hotter as it goes. It is useful to have a pump that can be controlled so that the rate of flow of the water is adjustable: in the summer it may be necessary to keep the water moving quite fast, to avoid overheating, while as it approaches winter the flow may be quite slow. In the coldest part of the year, there may be some days when very little heat is collected from the sun, and in this case it is necessary to have some arrangement to stop the circulation altogether otherwise heat will be wasted by pumping hot water from the storage tank out into the panel, where it will be chilled. Many systems have a thermostatic valve or control to the pump that only allows water to circulate if the temperature of the panel is higher than that of the storage tank in the house. This ensures that heat can only be taken in, not out.

You can obtain a list of the manufacturers of solar panels from the

Simplified diagram of heating system

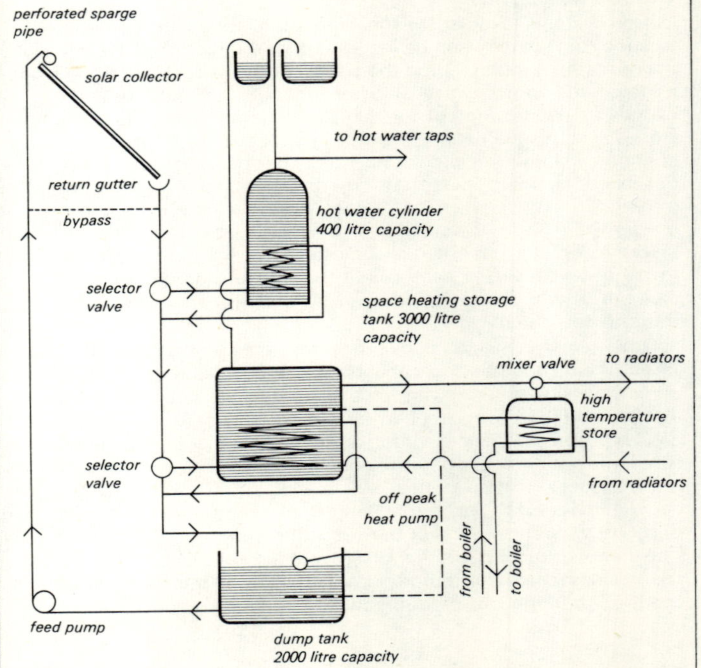

Collecting the heat from the Macclesfield solar roof. Water from the storage tank (bottom centre) circulates over the hot aluminium (top left) and heats water in a series of indirect cylinders. Conventional heating (bottom right) provides supplementary hot water

International Solar Energy Society (see page 77).

Homemade collectors

It is obvious from the above discussion that a closed-system solar heating plant is very much like a central-heating system, but in reverse. Both use panels with a large surface area to exchange heat between circulating water and their surroundings. In the case of the heating system these are designed to take in hot water and radiate heat to the room; in the solar panel they absorb heat and raise the temperature of the circulating water. The resemblance is so close that it is not surprising to find that many experimenters, concerned with self-sufficiency but with limited budgets, have used central-heating radiators as panels. It is necessary to strip off the light-coloured paint usually used on such radiators and repaint them with some matt black finish that will absorb heat. Such panels often work very well, although they may not look as elegant as a properly-made commercial model. The panelled type of radiator is better than the older design with deep recesses between the water-channels, as the sunlight is unable to penetrate to these recesses so that a good deal of the actual surface area is wasted as an absorber.

Siting a solar panel

Roof panels

The roof is a favoured place for solar panels, partly because it is usually high enough from the ground not to be obscured by the shadows of trees or buildings, so that the panel has a reasonable chance of catching the sun for as long as possible during the day. Roofs are also convenient because the pipe-runs to carry the water from the collector to the interior of the house can be kept fairly short and therefore do not lose much heat. However, perhaps the most cogent reason for mounting a panel on the roof, for most people, is that this is the only large uncluttered area in the house or garden which has any chance of getting much sun.

The simplest way to mount a panel on the roof is to set it up parallel to the slope of the rafters, just above the tiles, but this is not necessarily the best position for efficient absorption of solar energy. The aim is to mount the panel so that it receives as much of the sun's radiation as possible, especially in the winter months, and this means that the slope of the panel ought to be decided by considering the behaviour of the sun, as well as the constructional requirements of the roof.

First, the panel should if possible face due south. The actual path of the sun varies during the year (it is only at the equinoxes that it goes precisely from east to west), but on average the mid-point of its path lies in the south. Second, the angle of the panel to the horizontal should allow it to face the sun for as long as possible, particularly in the winter when the sun is low and days are short. The angle is complicated to calculate, but taking the average elevation of the sun at different seasons and different times of day, the absorption of heat is most efficient if the collector is set up at an angle of $L + 13\frac{1}{2}^{\circ}$ to the horizontal, where L is the latitude, in degrees, of the site. You can look this up on the Ordnance Survey or similar maps. Around Bedford, for example, at latitude 52°, the collector should be set up at $65\frac{1}{2}^{\circ}$: in Inverness, at latitude $57\frac{1}{2}^{\circ}$, the collector should be at 71° to level ground. These figures will ensure maximum efficiency during the winter months, when any heat from the sun is most welcome. They do not, of course, give such efficient collection in summer, but one of the common problems with solar collectors is that they tend to pick up too much heat during the summer months.

However, if your installation is only meant for use in the summer, for example in a holiday chalet, you can use a smaller collector set at the ideal angle for absorbing heat when the sun is high in the sky: an angle of $L - 23\frac{1}{2}^{\circ}$ is best under these conditions. These angles need not be constructed precisely to half a degree when setting up the solar roof or panel, but the closer you get to the figures quoted the better your collector will work.

One important consequence of these calculations is that it becomes obvious that the average house roof is not at the best angle for solar collection. Most of them have a slope in the region of $35^{\circ}-50^{\circ}$ to the horizontal, and many designs are even flatter than this. If your roof is also your collector, there is not much that you can do about this unless you have a steep south-facing side and a shallower northern side. Two very steep sides to the roof would involve using a great deal of expensive material in the construction. If however your collector is a separate panel or series of panels mounted on an existing roof it will pay to construct the supports so that the top end of the panel is set out more from the line of the roof than the bottom end. By using suitable metal struts you can arrange the panels at the correct angle.

Ground-level panels

Panels can also be mounted in the garden or on any unshaded area; many people use a row of collectors, set up at the right angle to the horizontal, along the south side of open-air swimming pools. These act both as solar collectors and as shade for the side of the pool on very

Solar panels mounted on a house wall near St. Moritz, Switzerland. Hot water is taken into the house through the pipes on the right

hot days. Water circulates through the panels in the normal way and is passed through a heat-exchanger to warm the pool water, giving a very welcome economy in heating costs. It would be possible, of course, to circulate the pool water directly through the panels, but as it is normally chlorinated and treated with chemicals to prevent algal growth, etc., there may be a risk of corrosion in the interior of the panels unless they have been specially designed for this work. Some PVC panels are resistant to chlorine, but the manufacturers should be consulted before using these with swimming pool water.

Where it is impossible to set up a collector facing due south, you have to settle for having it as near south-facing as possible. The efficiency will obviousy be less than ideal and the slope of the collector needs to be even steeper than the angles given by the calculation above, because the average elevation of the sun, as seen from the panel, will be lower. The exact angle for ideal performance depends on the site and the precise direction in which the collector faces, but, as a rough guide, add 1° to the slope of the collector for every 10° away from due south.

Precautions to avoid heat loss

Whatever the position of the collector, precautions must be taken to avoid loss of heat. Draughts can blow away a lot of warm air from the space between the glass and the actual collecting surface unless the joins between the glass and the frame are kept well-sealed and airtight. The pipework carrying water to and from the interior of the house must be lagged carefully, as should also any connecting pipes between separate panels.

To guard against frost, the water used in any circulating system should ideally be treated with an efficient anti-freeze such as ethylene glycol, especially when metal panels are in use, as these may be cracked by expanding ice if the system freezes up. Manufacturers of plastic panels often claim that these need no anti-freeze in the system because the plastic can expand, but even so it is better to take precautions against freezing as trouble may occur if the water outside the house freezes and blocks the movement of the interior sections—excessive scale formation may take place, for example, because the water is static.

Scale build-up is a problem in itself. A good deal of trouble is avoided because the circulating systems are indirect—i.e. the same lot of water circulates round and round, apart from occasional small additions to make up for losses—but a scale inhibitor should also be added to the water. In metal collectors, particularly aluminium, a corrosion inhibitor such as sodium nitrite or benzoate is a good precaution (this can usually be purchased mixed with the anti-freeze as 'inhibited' ethylene glycol, as used in car engines). Where the water is allowed contact with the air, as in 'open' systems like the Macclesfield solar roof, it is necessary to add an algicide (like those used for swimming pool water) to stop the growth of algae and moulds on the surface of the roof. This would not only be unsightly, but would cut down heat absorption.

How much heat does a panel collect?

The vital questions about solar collectors have not yet been answered: how much heat do they absorb from the sun, and can this be enough to make a house truly self-sufficient?

The answer to the first question can be worked out from the figures collected by a host of experimenters over the last few years, by measuring the actual amount of solar energy reaching us, and the efficiency of various designs of collector in converting this to useful heat (see for example, **Solar Energy for Man**, by B.J. Brinkworth). Obviously the amount of solar energy available depends on the time of year, the weather, the effects of the atmosphere, clouds, smoke and pollution in stopping it from reaching us, and the geographical position. However, the curves of the graph give a rough idea of the average solar power (in watts per square metre of collecting surface) actually detected in Plymouth, Kew and Stornoway, during the successive months of the year. Thus, taking a solar roof as in the Grants' house, about 45 square metres facing the sun, the total energy actually received from the sun would vary from an average of about 750 watts in December and January to over 10 kilowatts in June and July. Put another way, if all the solar power on such a roof could be converted to useful heat, it would be equivalent to about 21 kilowatt hours ('units') of electricity per day in the winter months, rising to 240 kilowatt hours per day in the summer.

This represents the maximum supply available: in practice the amount actually collected will be far lower because of the inefficiency of all the existing systems of collecting the energy. For example, most 'closed-system' panels can convert about half of the available solar energy into heat. Open systems may only be 30 per cent efficient (partly because the

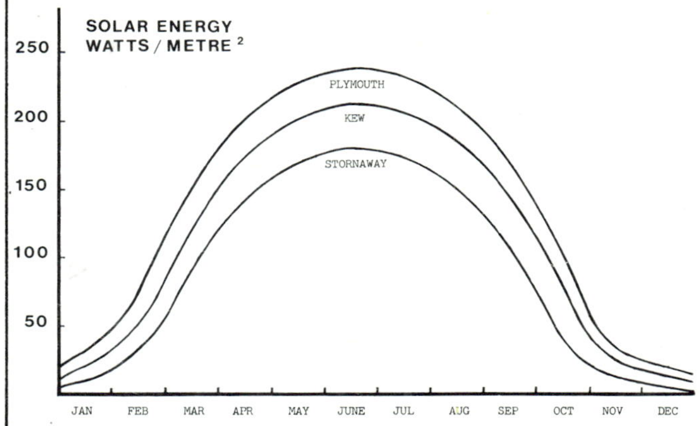

Average solar power (in watts/square metre) at various locations in Great Britain

water tends to produce condensation on the glass or plastic covering of the roof, and this interferes with the passing of the sun's rays). Similar losses are due to reflected radiation, dirt on the surfaces of the collector, failure to transfer all the heat from the surface of the collector to the circulating water and, most of all, the fact that as the water in the circulating system gets hotter, it begins to radiate heat outwards, and not all of this radiation can be stopped by the glass covering.

Apart from the intrinsic inefficiency of the systems, the angle of the collector to the sun has a profound effect. For example, if the roof of the Grants' house could have been set up at about 65°-67°, it would have collected more heat, particularly during the winter, than it will at its actual angle of 34°.

In the final analysis, therefore, we have to accept that the heat absorbed by a collector with an area of say, 50 square metres is not likely to be more than the equivalent of 10 kilowatt hours per day in the coldest part of winter, rising to about 100 kilowatt hours per day in the summer. The actual figures may be even lower than these if the efficiency of the collector is less than about 50 per cent, if the angle of elevation at which it is mounted is not the best for the site or when the installation is in the far north of Britain.

This really provides an answer to our second question—whether solar power can make a house self-sufficient for heat energy. If we are thinking of a house constructed and managed in our present wasteful style, the answer clearly is no—the sort of house that would have a roof area on the south side of 50 square metres would probably use about 70-80 kilowatt hours per day for heating and cooking in winter, and the sun can clearly only provide a fraction of that amount. Some saving in fuel costs could obviously be achieved, but the ideal of self-sufficiency would be a long way off.

On the other hand, if we are prepared to make changes not only in our means for supplying energy, but also in the methods we use to deploy it, the 10 kilowatt hours per day from a solar heater could be made to provide a much larger fraction of our heating needs. To begin with, this would mean adopting an entirely new set of standards for insulation, compared with the relatively low standards we tolerate at present. That great improvements can be made is shown by experiences in various experimental self-sufficient houses, including one in Sweden—a 'house for the future' indeed—that uses heating only for the cooker and hot water for washing. There is no separate 'heating system' at all. However, the insulation systems are so effective that the waste heat from these household activities is quite enough to keep the house comfortably warm even in the Swedish winter.

Apart from insulation, the best alternative way to improve the effectiveness of solar heating is to have some arrangement to store some of the heat absorbed during the hottest periods. In our example above, for instance, based on a house the same size as the Grants', the daily yield of heat in June and July averages about 100 kilowatt hours—an embarrassingly large amount of energy to use up in summer even if the family take endless hot baths, wash all their clothes every day and cook far more extensively than most families do in summer. If only a small fraction of this heat could be stored for the coldest winter months, the problems of self-sufficiency could be solved.

Storage of solar heat

Apart from the long-term advantages of storing heat from summer to winter, every solar-powered system needs some means of storage to allow for the cloudy or foggy days, when the sun is scarcely visible, and for keeping the house warm at night. At present the only effective way of storing such energy is to make a large weight of material hot when the power is abundant and extract heat from this when necessary. This is the principle of the night-storage heater, where the filling of the radiator is usually a heavy mass of fire-brick or drums full of granular mineral material. Water is an even more efficient medium for storing heat, because it takes up more energy per degree rise in temperature than most other substances.

Even so, the amount of storage material needed to make any substantial heat reservoir is large, many times more than the familiar 273-litre storage tank for hot water in most houses. Taking our figure of 80 kilowatt hours as the average winter daily requirement for a medium-sized house (which would mean a comfortable degree of warmth), we need to heat 1,122 litres of water from 15°-82°C to retain one day's requirements of energy. To keep a month's supply, say for January, we should need a tank containing about 31,000 litres of water, weighing about 30 tonnes.

Not many houses can accommodate such a tank inside, although it is quite possible to set it up outside, with proper insulation. The Macclesfield house incorporates tanks containing about 6,000 litres, which would give two days supply of heat and two tanks are set up in the service area next to the house (mainly so that the heat does not get oppressive in the summer). A design for a self-sufficient house developed by Alex Pike and his colleagues in the Department of Architecture, Cambridge, goes further than this, with a kind of basement entirely filled with water which is circulated over a solar roof. The tank in this case holds 45,000 litres. However, this project has not yet been completed, so no one knows for sure how well the system will work. There could be complications in the summer if the underfloor area gets uncomfortably hot from the water tank, although of course, if it were installed with a ventilated gap it would probably be satisfactory.

Possibly the simplest storage idea of this kind is that developed by an American enthusiast for self-sufficiency, Steve Baer of Albuquerque, New Mexico, where the days are very hot and the nights very cold. Most conventional houses there use a great deal of energy all the time, running air-conditioning plants during the day and electric heaters at night. Baer's house has the south wall infilled with a stack of oil drums filled with water. During the day they absorb the sun's heat and prevent it from getting into the house excessively. Just before nightfall, an insulated shutter is lowered over the outside faces of the drums, so that the heat they have collected during the day is radiated into the house throughout the night. Thus a very simple system, using no energy except solar heat in a particular location, can take the place of a sophisticated and energy-consuming central-heating and air-conditioning plant.

A similar 'primitive' but effective plan has been used to keep greenhouses warm at night. A large pit is dug in the floor of the greenhouse and filled with pieces of size graded stone, selected to allow plenty of circulation space for air to pass among them. During the day a fan draws the hot air from the roof area of the greenhouse through the pit, so that the stones are heated up. At night the fan is reversed, and the warmth is circulated round the plants. Such a system uses no power except a small amount of electricity to run the fan, while giving an equable climate in the greenhouse during much of the year such as could only otherwise be obtained with an expensive heating plant. There are possible condensation problems—the pit should be drainable.

Such a system has also been proposed for houses: a long air duct runs from the loft area right through the centre of the house to an underfloor area filled with a suitable heat-storage material. A fan at the bottom of the duct sucks down hot air from the roof-space when the sun is shining and passes it over the heat-storage reservoir. At night the fan can be switched off, so that the store heats the house from below, or it can be reversed and hot air blown from the store into the rooms of the house through other ducts.

Conservatory heating system (rock store)

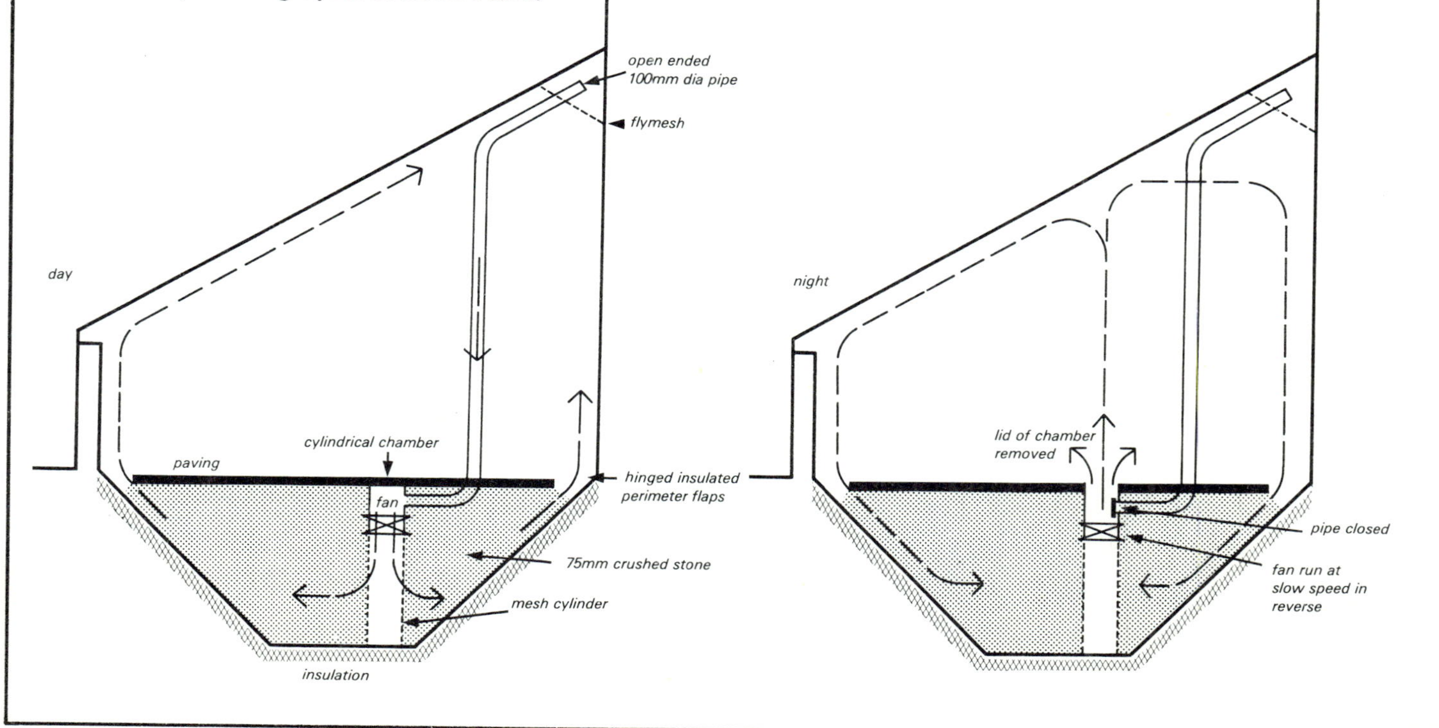

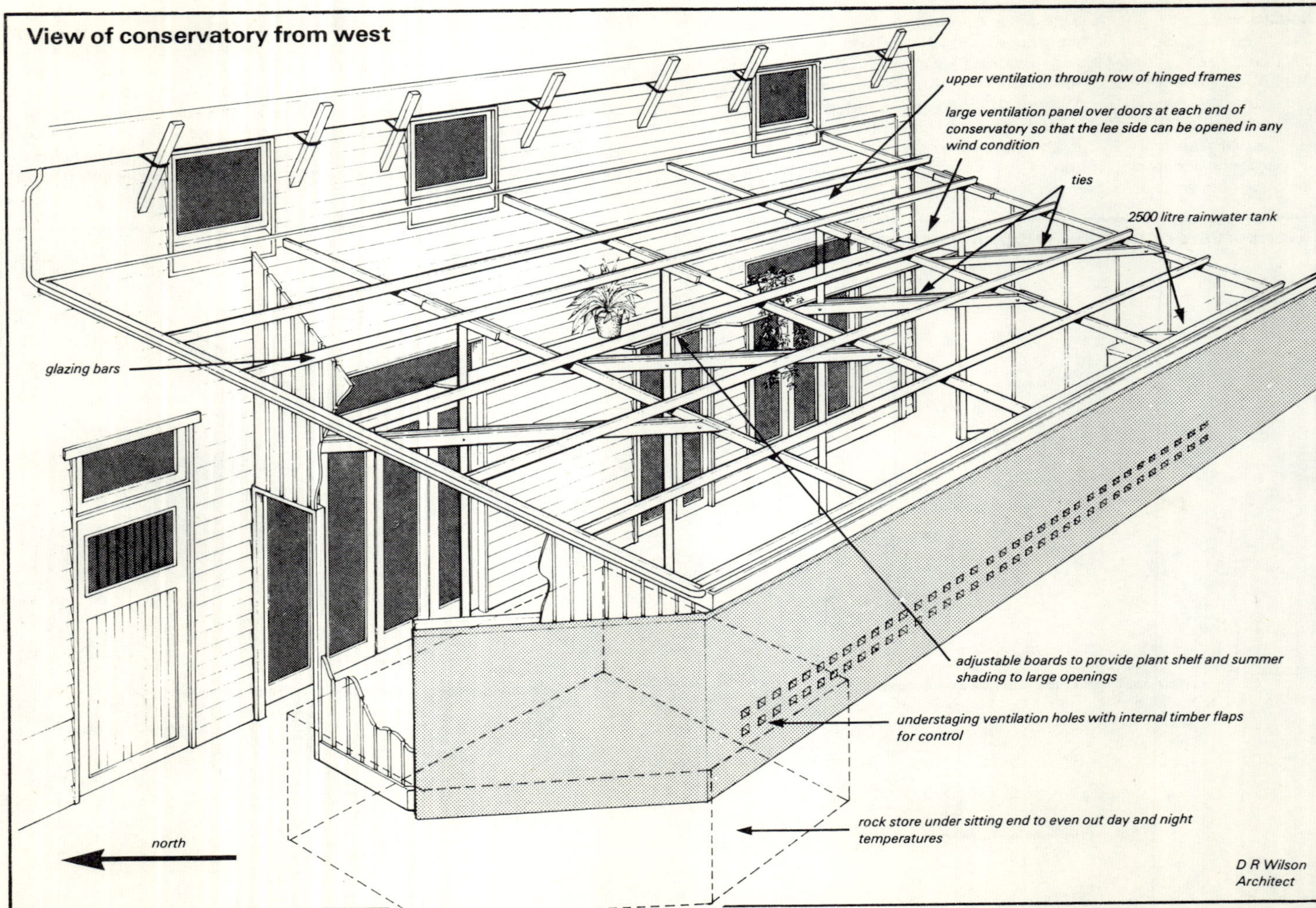

View of conservatory from west
upper ventilation through row of hinged frames
large ventilation panel over doors at each end of conservatory so that the lee side can be opened in any wind condition
ties
2500 litre rainwater tank
glazing bars
adjustable boards to provide plant shelf and summer shading to large openings
understaging ventilation holes with internal timber flaps for control
rock store under sitting end to even out day and night temperatures
north
D R Wilson
Architect

A vent under the bedroom window allows the Grants to use warm air collected in the greenhouse, or to ventilate their bedroom without draughts

Many designs for self-sufficient houses incorporate very massive walls on the south side, so that the structure itself acts as a 'storage heater'. For instance, in the Odeillo valley in the French Pyrenees, a group of scientists connected with the giant solar furnace recently built there have designed and built a group of houses with south walls made up of two very thick layers of masonry with a cavity between. Vents at the top and bottom of the walls can be set to allow air to pass in and out of the house, connecting the cavity either with the rooms inside or the open air. During the winter, the outer wall is heated when the sun is shining and warm air builds up in the cavity. A vent is opened to allow this to circulate into the house. During the summer, when the Pyrenean sun beats down uncomfortably, the warm air from the cavity is passed out to

A solar house in the French Pyrenees uses the whole south wall as a collector

The reflectors of the solar furnace at Odeillo in the French Pyrenees. The central furnace at the focus can develop 6,000°F

the open air and cooler air from the north side of the house comes in to take its place, thus producing a cool air-flow.

However, long-term storage of heat from one *season* to another is still difficult, unless you can design your house with facilities to keep about 50 tonnes of water in storage. Several ideas have been put forward for converting solar power into a more convenient form—for example, using it to decompose water into hydrogen and oxygen and then burning the hydrogen as fuel during the winter—but so far none of these ideas have been put into practice except as small-scale experiments. Several people have said, in fact, that the most efficient way to store solar heat at present is to use it to grow trees, and then cut them down and burn them during the cold seasons.

Solar cooling

It may seem odd to think of using the heat of the sun to cool buildings, but in many ways this is the best way to make use of the sometimes excessive heat collected by solar systems in the summer months. Long-term storage from summer to winter is difficult (see page 18), while cooling needs only the application of well-known principles of refrigeration and air-conditioning. While there are not yet 'off-the-shelf' cooling systems, most of the problems have been solved and the free supply of summer heat can be utilized in the most direct way to make the home more tolerable.

The principle is exactly the same as that of the domestic refrigerator. One end of the system compresses a fluid, which gets hot in consequence (much as a bicycle pump gets hot as you compress the air in it) and the heat dissipates itself in a radiating coil. The fluid then expands at the other end of the system and in doing so it gets cold and extracts heat from anything round it. The fluid then returns to the compressor, and the cycle begins again. In the refrigerator we use gas or electric power to compress the fluid and the heat dissipates itself in the kitchen (through the coils at the back of the refrigerator) and the interior of the cabinet is cooled by the expansion end of the system. In a solar cooler, power from the sun is used to compress the fluid, and the heat produced is allowed to escape to the open air. The cold end of the system consists of a set of coils inside the house, usually with a fan to distribute the cool air round them throughout the house.

Conventional refrigerators can be run either with an electric pump ('compressor' type) or from a small electric or gas heater which compresses the fluid by the force of expansion ('absorption' type). Solar cooling devices can also be of these two varieties. Compression coolers are run from a solar-heated engine (usually a Stirling engine—a type of power unit invented by a nineteenth-century clergyman which has only recently become important because it works from almost any source of heat and is entirely non-polluting when run properly). The absorption type can be run from the hot water developed by a solar panel or by focusing mirrors on the 'hot end' of the cooler system and collecting the sun's heat this way. In the latter case the cooler does not run at night, but this is often unnecessary anyway. The mirror (or 'concentrator') system gives the highest efficiency of all these three methods.

So far no commercial solar heater/cooler systems are on the market. The National Science Federation in the United States are spending large research funds on developing such a system, which is, of course, far more necessary in most of the United States than it is in Britain. Some independent researchers in this country are also at work on designs.

Solar batteries

When satellites and similar craft are moving around in the almost empty spaces between the planets, they are cut off from most conventional sources of energy. Yet they have to be able to transmit information back to earth by radio or TV, operate controls and instruments, and do a number of other things that need power. Obviously they can carry batteries, but sufficient batteries for the months that some of the craft spend in space would be very heavy and bulky.

The secret of their power supply lies in those curious flaps, rather like wings, that they spread out as soon as they enter space. These are arrays of solar cells, which take in sunlight and convert it to electricity. Some spacecraft, such as the communications satellite *Intelsat 4*, may be generating as much as 900 watts continuously from sunlight.

These cells are now available for terrestrial uses, and many of them are already employed for providing electricity in situations where conventional power supplies are hard to obtain—beacons for use at sea or on desolate rocks, fire warning devices that can be set up deep in the forest, power for radio and telephone links in places like central Australia, etc.

The heart of a solar cell is usually a very thin slice of pure silicon which has been treated ('doped') with very small amounts of phosphorus and boron. The manufacturing technique is very similar to that for making transistors or integrated circuits for other electronics work and requires the same scrupulous cleanliness and control over the conditions. There is one major difference, however. Conventional electronics parts are steadily becoming smaller and smaller, as 'miniaturized' equipment is superseded by 'microminiaturized' devices and increasing numbers of components are squeezed on to a tiny chip of silicon, as in the complex circuits for calculators or electronic watches. On the other hand, solar cells have to be as large as possible to pick up sufficient sunlight and this means that the manufacturers have to develop entirely new methods for making silicon chips that are thousands of times larger than the average transistor's 'heart'. This is why solar cells are so expensive at the moment, but a manufacturing breakthrough could reduce the price at any time.

A typical commercial solar battery consists of 30 individual silicon cells mounted on a flat base 450mm x 345mm. This panel will generate electricity at a little over 12 volts in normal daylight and in an average summer week could develop about 300 watt-hours, less than a third of a 'unit'. As these panels cost about £150 at present, the idea of using them for the supply of domestic electricity is not promising: most people buy them to charge batteries for a boat or caravan, or similar situations where mains power is not available.

The panels are very robust, standing up to salt water, motor fumes and extremes of heat and cold, so they need very little attention and would be ideal for mounting on a roof, if they could be made more cheaply.

A group of 5 Lucas solar cells make an emergency battery charger

Chapter 3
Wind Power

Ever since the first man put up a bark sail to help his boat along, people have been using the energy of the wind to save their own muscle power. No one knows when the earliest sailing boat was built, but windmills for pumping water and grinding corn have been in use for at least two thousand years.

The simplest type of windmill has a series of sails made of canvas or similar material fastened to the spokes of a kind of wooden wheel. The canvas is set at an angle to the wind, like the sail of a yacht, and the wind rotates the wheel, which can be used to lift water from the river to the fields or to turn a simple grindstone. Such mills are still widely used in India, China, the Middle East and even the more remote parts of Spain and Portugal. Apart from being very easy to make and maintain, they have several advantages in aerodynamic design that have only recently been recognized by modern wind-power researchers.

The traditional European windmill, with its rotating cross of wooden arms, was also originally made to have stretched canvas to catch the wind. Millers had a busy time climbing out on the sails to spread out more canvas as the wind dropped or reef it in as the wind got stronger, exactly like the crew of a sailing ship. As a large mill could easily generate 70-80 horse-power in a brisk breeze, controlling the speed of the sails was no easy task. Some of the most resourceful inventors of the eighteenth and nineteenth centuries applied themselves to the improvement of windmill mechanisms, and by the time they were superseded by the steam engine, windmills had become full of highly sophisticated machinery. Now that the fuel for our steam engines and their modern equivalents (petrol, diesel or gas engines) is becoming increasingly scarce and expensive, many people are looking again at the wind as a source of natural, non-polluting and, above all, free and inexhaustible energy.

Of the developed countries, Great Britain, in company with Holland and Denmark, is one of the most windy, with a reasonably constant supply — there are very few days per year when the air is so still that some kind of windmill cannot be operated. In addition, most of our strong winds occur in the winter, when solar power is at its least effective, so the two systems are complementary. And on average, about three-quarters of the total wind energy we receive every year comes

A 200-kilowatt wind-powered generator built by the Danish State Electricity Board at Gedser

between November and April.

The scale of the supply is vast—if we could harness all the wind energy available just over the west coasts of Scotland, Wales and England we could provide between a quarter and a third of the electricity requirements for the whole country. On a worldwide basis, the total wind energy over the land surface alone is equal to about twice the output of all the power stations now in existence. Yet at present we use only a tiny fraction of this enormous reservoir of free power.

Using the wind

The nature of wind makes it ideal for driving machinery—driving pumps, millstones, mechanical hammers, etc. in the past; electrical generators now. In Denmark, where other sources of energy such as coal, oil or water-power have always been in short supply, wind-powered electricity generators have been in use since about 1900, including some very large windmills set up on tall concrete towers and connected to the Danish national grid. In Britain and the United States many remote farms managed for years with simple wind-generators as their only source of electricity.

The amount of electricity, or any other form of power, that you can extract from the wind depends on the size and efficiency of your wind sails, and particularly on the diameter of the circle they can sweep out. Obviously the bigger the circle, the more wind they can catch. The power also depends, even more critically, on the wind speed, rising sharply as the speed increases. For example, a 32-kilometres an hour wind can give 8 times as much power as a 16-kilometres an hour one, and a 48-kilometre an hour wind 27 times as much. This makes it very important to choose a spot for your generator where average wind speeds are high and there are no obstructions to slow down the flow of air, and it also means that designs for both the actual windmills and the electrical generators have to be able to cope with an enormous variation in load between the lowest wind speeds and the highest ones. The table below shows the sort of output you can expect from a wind-generator of average efficiency at different wind speeds.

Windmill outputs in watts (W) and kilowatts (kW) at various wind speeds

Windmill diameter		Wind speed							
		kilometres per hour (miles per hour)							
(feet)	metres	8 (5)	16(10) (Light breeze)	24 (15) (Moderate wind)	32 (20)	40(25)	48 (30) (Strong wind)	56 (35)	64 (40) (Gale)
(1)	0.3					20W	30W	50W	70W
(2)	0.6		(Under 10 watts)	15W	35W	70W	120W	190W	280W
(3)	0.9			30W	80W	150W	270W	420W	630W
(4)	1.2		20W	60W	140W	270W	470W	750W	1kW
(5)	1.5		30W	90W	220W	430W	740W	1.2kW	1.8kW
(8)	2.4		70W	240W	560W	1kW	2kW	3kW	4.5kW
(10)	3.0	15W	110W	370W	880W	1.7kW	3kW	5kW	7kW
(15)	4.5	30W	250W	830W	2kW	4kW	7kW	11kW	16kW
(20)	6.0	60W	440W	1.5kW	3.5kW	7kW	12kW	19kW	28kW

This underlines the tremendous effect of wind speed on output. In an 8-kilometre an hour breeze even a 6-metre diameter mill can produce barely enough power to light one small electric light bulb, yet in a gale it could theoretically furnish enough for the most extravagant requirements of two average all-electric households. There is obviously no point, however, in fitting up a large, heavy and expensive 28-kilowatt generator to a windmill if it is only going to work at full capacity for a few days a year. On the other hand, if you use a smaller generator, you must provide some means of slowing down the windmill when gale-force winds actually arrive or your generator will burn out from overloading.

In addition, as may well be imagined, a 6-metre diameter windmill in a gale is an awe-inspiring hazard to lives and property unless it is properly designed to stand up to the stresses and strains, and the braking mechanism is efficient. Most mills of any size are designed so that their running speed can be adjusted, or even self-adjustable, as the wind varies.

Other factors to be considered are the 'dead' and calm days, when there is little or no wind. No design of windmill can operate in a flat calm, but some patterns are better at making use of very light winds than others and will keep ticking over while other designs are immobile. The cut-in speed'—the minimum wind speed to ensure movement of the mill—is an important property of every mill design.

All these factors make the construction and siting of wind-generators a highly skilled operation, especially with the larger generators.

The best compromise is to assess the *average* wind speed for your site and use this as a basis for deciding on the size of windmill and generator that you can reasonably run. On the west coast, for example, you can probably rely upon average speeds of 24-kilometres per hour or even more for most of the year: in the Midlands you may expect only 16-kilometres per hour. As the highest wind speeds and the greatest demand come together in the winter, it may be best to take your average over the period from October to March only. Fortunately, unlike sunshine figures, the wind pattern seems to stay remarkably constant from year to year for any particular site.

For accurate measurements you should buy or hire an anemometer, (from Conservation Tools and Technology Ltd., for example—see p 00) a simple device that spins round in the wind and operates a counter. By taking readings from the counter at regular intervals you can log the wind speed pattern and compile a complete dossier on wind behaviour on your site (what the experts call the 'wind regime'). Alternatively you can ask your local meteorological office for their wind data, but while these are of general assistance for describing the wind patterns on the large scale, your own very local conditions can be significantly different.

In the absence of an anemometer, you can make quite a good assessment of your local wind pattern by looking for these familiar signs:

Wind speed—kilometres per hour	
Under 1.5	Smoke rises vertically from chimneys. Flags limp against the flagpole.
1.5-5	Smoke drifts slowly.
6-11	Breeze distinct on face. Leaves on trees rustle.
13-19	Leaves and twigs on trees in constant motion.
21-29	Dust clouds raised. Small branches of trees sway.
31-39	Small trees in leaf begin to sway or bend.
40-50	Large branches of trees move. Audible whistling in telephone and other overhead wires.
51-61	Large trees sway. Walking against wind difficult. Howling around buildings and roaring as wind goes through large trees.
63-74	Twigs break off trees. Difficult to stand up straight in the wind.

If you are on top of a hill you should have little difficulty in finding a suitable site for a windmill, and even at the bottom of a hill matters are not usually too bad. The worst position is halfway up a hill on the leeward side, as the wind tends to be deflected upwards by the hilltop and passes over the windmill. Local obstructions also play a large part in deciding the wind regime in a particular spot. A tall tree or a house, for example, can cause turbulence, which puts extra strain on the windmill without increasing its output.

If there is a prevailing wind direction you can minimise this effect by setting up the wind-generator on the windward side of the obstruction. If this is impossible, or if the wind is constantly varying in direction, you must either place the generator high up on a mast or tower, preferably at least 4.5 metres above the top of the obstruction, or make sure that it is at least 365 metres away. In any case a high mast is an advantage because wind speeds are usually higher about 9 metres up than they are at ground level.

Such masts and towers have to withstand very high pressures from the wind—an 80-kilometre an hour wind can exert over 0.5 kilonewtons per square metre pressure, tending to push the structure sideways—and there is also strain caused by the movement and vibration of the windmill itself. Sudden changes in the wind direction cause more strain than a steady high wind, as the mill swings round to face the wind and alters the load on the supports. Even a small generator exerts far more stress than a static structure such as a TV aerial.

For this reason, you need to stabilise your windmill very carefully. For small mills, up to about 1200 millimetres diameter, you can do this job yourself if you follow the advice given on page 71, but if you are contemplating buying a large commercial generator, make sure that the suppliers can also provide a properly engineered mast or pylon suitable for the machine, with adequate stays or guy-ropes to brace it against the wind.

Types of windmill

Many different designs of mill have been invented to take advantage of the energy in the wind, and each type has its own peculiar advantages and disadvantages. They can be conveniently classified into two main types—those with sails more or less vertical, and a horizontal shaft, and those with vertical shafts. The latter design may not seem very familiar, but in fact the spinning advertising signs outside many petrol stations are simple vertical-shaft windmills.

Of the horizontal-shaft mills, the oldest and most numerous up to now are the *slat mills*, with 4 to 8 sails made up of a wooden framework filled in with canvas or shutters. While many of these are still operating very successfully, the design is not favoured for electrical generators, partly because the woodworking skills necessary for constructing such mills are now very rare, but mainly because the slat mills tend to turn very slowly and ponderously, even in a brisk wind. This does not matter for pumping water or grinding corn, or any of the other traditional tasks of the windmill, but alternators and similar electric machines work best at fairly high speeds of revolution. To convert the motion of a slat mill to the 700 or so revolutions per minute needed for many generators, the shaft speed must be increased by a gearbox or pulleys and belts, and this adds to the weight, cost, and frictional losses of the windmill.

Although Danish engineers have made a few slat-type wind-generators, using gearing to increase the shaft speed, the most popular solution to the problem is to have a blade shaped like an aeroplane propeller, with 2, 3, or 4 blades. This develops no more power than a slat-mill of the same diameter, but it turns much faster at any given wind speed. Usually the speed of the tips of the blades of a well-designed propeller is about 6 times the wind speed, so that a 180 centimetre diameter propeller in a 40 kilometre per hour wind will have a tip speed of 240 kilometre per hour and turn at about 700 revolutions per minute, which is suitable for many generators without any intermediate gearing.

Some very large propellers have been constructed for electricity

Canvas-sailed windmill near Peniche, Portugal

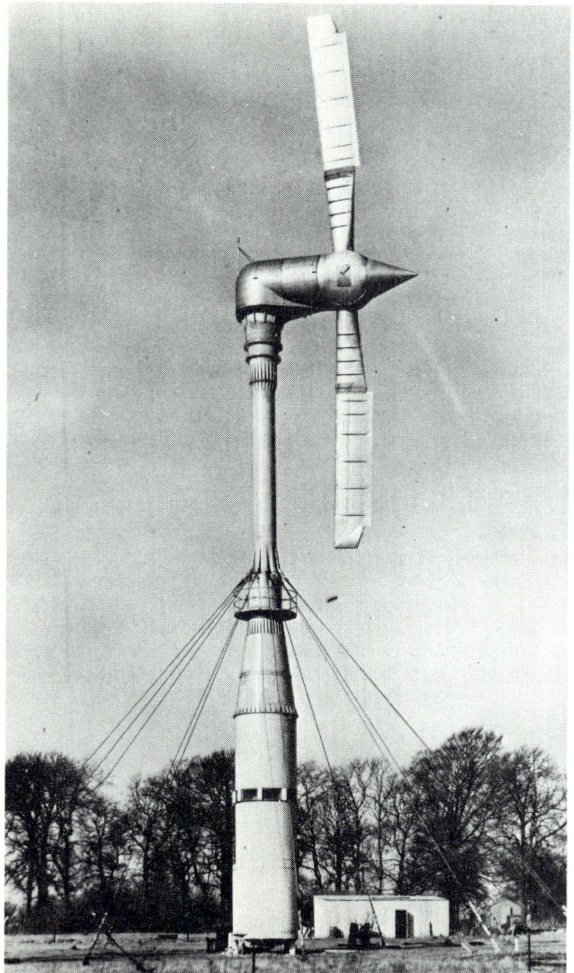

A traditional post-mill at Pitstone, Bedfordshire

Experimental 100-kilowatt wind-powered generator built by the Electrical Research Association

generation, the most impressive so far being a 53 metre stainless steel blade mounted on a 34 metre steel tower on a windswept hill in central Vermont, Grandpa's Knob. This giant ran a generator capable of producing 1.5 megawatts (1megawatt = 1 million watts), enough to provide electricity for 300 homes.

Unfortunately, these very large generators have had a sad history of failure in very high winds. The Vermont propeller broke apart in a tornado, hurling one of its blades almost a kilometre, and similar structures have repeated the story, either the blades or the actual supporting tower collapsing under the strain of freak winds. It seems ironic that all the resources of modern engineering and materials have not yet produced a satisfactory high-power windmill, while the old wooden-sailed mills kept going for centuries with no more attention than could be given them by the miller himself and the local carpenter.

One of the major problems with propellers is that they are difficult to adjust to varying wind speeds. Obviously the area of the blades cannot be altered, as with a slat mill, and the only solution is to change the angle of the blade to the wind, i.e., the pitch. Variable pitch propellers are of course used in aircraft, but the mechanisms are rather complex and add to the engineering problems of designing large propeller windmills.

Smaller propellers, suitable for driving a generator with power capacity up to about 6 kilowatts, are generally available on the market, some with fixed-blade systems and a few with variable pitch ('self-feathering blades'). These can supply most of the electricity requirements for a family or a small farm. If you want to run a wind-generator just to provide part of your electricity requirements, or for standby in emergencies such as power cuts, it is quite feasible to construct a home-made wind-generator with a propeller made of wood or fibreglass driving an alternator from a car or lorry—you will find more about this on pages 70-74.

Another type of mill, still to be seen in many country districts, is the multi-blade type or windwheel. These turn even more slowly than the slat-type mill, and are usually used for jobs like pumping water from ditches, but they can be adapted for electricity generation by gearing up the shaft speed. One of the advantages of the windwheel in many sites is that its cut-in speed is very low, so that it keeps going even in the slightest breeze. This makes the type very useful for such work as charging batteries for a boat or caravan, or for running a remote signal light or radio link.

The simple canvas windmill mentioned at the beginning of this section is a type of multi-blade mill, and it has only recently been recognized that this apparently crude machine has a very efficient built-in safety device. With a metal windwheel there is no easy way to change either the area of the blades or their pitch to the wind, so when high winds occur it is usually necessary to tie down the wheel or otherwise immobilize it. The canvas sails, on the other hand, tend to flap when the wind gets too strong and the mill is in danger of running out of control. This flapping cuts down their efficiency in collecting the wind (they 'spill the wind') so the windmill automatically slows down again to a reasonable speed.

Yawing

All horizontal-shaft windmills present one difficulty in their construction. To run at all efficiently, the sails have to face into the wind at all times. This means that unless the prevailing wind is remarkably constant in direction, the mechanism has to be mounted in such a way that the sails can be turned to face any direction. This process is called 'luffing' or sometimes 'yawing'—both words another reminder of the close links between the traditional windmill and the sailing ship.

Most wind-generators have the mill and generator mounted on a central pivot, and a tail extended out downwind to turn the mill into the wind. Some mills have their sails fixed on the downwind side so that the pressure of the wind automatically turns them in the right direction.

However, when the generator is working, there must be some way to carry the electricity from the movable upper section to the fixed mast or support. Long loose cables can be used for small installations—they may wind themselves round the mast from time to time, but if the mill is small they can be unwound by hand. The only really efficient connection for large generators is to have sliding contacts ('slip-rings') between the generator and the fixed mounting. These slip-rings must be designed so that they stand up well to weather and stress, for any corrosion or dirt on the sliding surfaces will interfere with the flow of current and cause loss of power.

Vertical-shaft mills

One way to eliminate all the problems of yawing and slip-rings is to build the windmill with sails spinning around horizontally, like a turnstile, and mounted on a vertical shaft. This will mean that whatever the direction of the wind, one or other of the sails will be at an angle to be moved forward by it and the mill will start turning. As the shaft is always in the same position, the generator or any other mechanism to be driven can be fixed at the end of the shaft and no moving connections or slip-rings are needed.

However, some vertical-shaft mills suffer from a disadvantage. As one sail moves forward with the wind, the opposite one will be coming back against the wind, and unless something is done to make the wind-pressures unequal on the two sails, the mill cannot move. Various ingenious devices have been suggested to get over this difficulty. Robert Beatson, in 1798, invented a vertical-shaft windmill with sails rather like a Venetian blind or louvre: when a sail was travelling with the wind, the slats of the louvre closed and caught the full pressure, but as the sail came round in the opposite direction, the slats opened so that there was very little resistance to the air-flow.

Some modern vertical-shaft mills depend on shaping the sails so that the side travelling with the wind catches the full pressure, but that going against the wind is convex, or otherwise curved, so that the air-flow slips round it and does not oppose its movement. One of the simplest windmills of this kind was invented by the Finnish engineer Sigurd Savonius, and is called the Savonius rotor. It consists of two hollow half-cylinders (in simple mills these are usually made from an oil-drum sliced down the middle) arranged on opposite sides of a rotating platform. When the wind blows, the concave interior of one of the 'sails' catches all the air-pressure and moves forward. The opposite 'sail' has its convex face to the wind, and as the air-flow moves over it there is a slight forward suction (an aerodynamic principle known as the Magnus effect) so that it is actually drawn forward *against* the wind. Thus both halves of the windmill are trying to rotate in the same direction and the efficiency is far higher than that of any simple, flat-bladed, vertical-shaft design where one half of the rotor has to be shielded from the wind for it to work at all.

A Savonius rotor

Another kind of vertical-axis mill dated back to the 1920's. Designed by a French engineer, G.J.M. Darrieus, it has thin, aerofoil section blades formed into the kind of curve made by a skipping rope. These blades give it a shape not unlike the outline of a rugby ball standing on end. It is hard to imagine that it would work at all, since there is no bucket shape to catch the wind as there is in the Savonius mill. Instead it depends on generating 'lift' in the same way as an aircraft wing: stand that wing on end and it will generate horizontal rather than vertical lift as the air flows over the aerofoil section; tether it and it will move in a circle round the tethering point.

The Darrieus mill usually has three narrow wing-like blades and once it is revolving it will turn at higher speeds than horizontal axis machines. It does suffer from one disadvantage in that it needs to be set in motion before it will produce lift. One solution has been to fit small Savonius rotors onto the vertical shaft so that they set the device turning. Once the main blades reach a certain speed in relation to the air flow they take over and the revolutions rapidly accelerate.

The combination of its high revs and the fact that the vertical shaft *has* to come down to ground level means that many of the gearing problems are reduced and that there are fewer problems in transmitting the shaft power to where it can be used.

Furthermore, like all vertical-axis mills, it needs no tail to bring it into the wind—it can take whatever wind from whatever direction. All this adds up to efficiency at low cost.

Other inventors are experimenting with windmills where the vertical 'sails' are shaped rather like helicopter blades, and one Canadian design has been claimed to be able to generate about 1 kilowatt in a 21-kilometre per hour wind, with a very simple and cheap construction. Dr. Peter Musgrove of Reading University, however, has simplified the design still further so that the blades are an untapered aerofoil section without 'twist' and incorporate a simple but effective feathering device.

Peter Musgrove's design for a vertical-shaft wind-generator being tested at Reading University

The future of wind power

With so much experimentation going on, and so much past experience to be drawn upon, it is natural to ask why more wind-powered equipment is not already in use.

One of the answers is the enormous variability of the source of energy—the wind. Solar power is also variable, between summer and winter, bright days and cloudy ones, but even during the dullest and coldest periods it is usually possible to extract some power from the sun during the course of each day. With the wind, however, there are some days when it furnishes no power at all.

At the other end of the scale, the extremes of solar power in midsummer may make everything uncomfortably hot, but are unlikely to do worse, while maximum wind power, in the form of a gale or tornado, is a terrifyingly destructive force. Planning for these extremes of wind power has been one of the biggest headaches for designers.

Days of dead calm can be allowed for by storing the electrical energy in batteries. Storage batteries are made with capacities up to 180,000 watt-hours and a life of about 8-10 years, which means that you could draw a steady 2 kilowatts from such a battery, day and night, for nearly four days before it needed charging. Perhaps, as a great deal of the electricity used in the modern house is for heating, it would be more sensible to use solar power directly as heat, and wind power for furnishing electricity for all the ancillary services, which would mean that far less than 2 kilowatts would be needed and a smaller battery would suffice. Solar panels usually need pumps, thermostats, electronic control gear for the more sophisticated systems, but none of these take large quantities of electric power. The running of the house would require electricity for lighting, TV, radio, deep-freeze, refrigerator, the motor of the washing machine (but not the electric heating element—hot water would come from the solar heating system), electric razor, sewing machine and power tools, but again this type of equipment does not take a fraction of the power consumed by electric heating, and an average supply of about 1-1.5 kilowatts ought to be quite adequate.

Storage battery capacity to give, say, 1 kilowatt average for 2 days continuous use would be 48 kilowatt-hours. This could be set up as one large stationary battery, but it would be cheaper, because they are mass-produced, to use a number of car or commercial vehicle batteries connected together. The 24-volt batteries made for lorries are often made with a capacity of about 5 kilowatt-hours, so ten of these would give a perfectly adequate electricity supply when connected to a suitable wind-generator, and even out the variations in the supply. This obviously represents a considerable capital outlay, but the return on it can be measured by the savings in mains electricity over years to come.

The future of very large windmills to provide electricity for the public supply is more hazy than that of 'domestic' generators. The big mills have obvious attractions, apart from the use of free energy. Unlike the smaller generators, which have to be positioned near the house they are to serve, public windmills can be sited where the wind condition is most favourable, even if this happens to be the top of an isolated hill, so their energy conversion is likely to be very much more efficient. Being large and high by necessity, they can make use of the fact that wind-speeds are higher above 9 metres from the ground, and there is a lot of power available to a windmill with, say, a sail-circle that gets to 60 metres at its highest point. Such regions are not accessible to the domestic windmill, unless we are reconciled to a forest of 30-metre pylons in our back gardens. If we had enough large generators, they would do much to even out the variability in the wind supply at any one time, because it would be unlikely for the weather to be calm all the way from Devon to Lancashire, for example. Economically, although such a chain of giant windmills would be costly to set up, several expert analyses have shown that the electrical output, in proportion to the capital cost, would be at least as good as that from building nuclear power stations, without any of the corresponding risks. A nuclear accident can easily involve hundreds of square kilometres with insidious radiation poisoning, and affect the lives of thousands of people. A windmill, even a large one, breaking up will only cause damage within a few hundred metres.

Sad to say, it is mostly *outside* the UK that any real state investment has been made into developing wind-power. The research into the Darrieus vertical-axis mill has mostly been carried out in Canada while in America NASA has constructed a 38-metre diameter horizontal-axis mill that will produce 100kW—enough to supply 25 homes—as an initial move to investigate the possibilities.

In the UK the initiative has come more from individuals, universities and commerce. In Surrey, the Wind Energy Supply Company has started with the small and medium-sized units of their building programme—6 metres and 18 metres diameter respectively. The smaller is intended for domestic use; the larger for heating commercial greenhouses. Both are based on the adaptation of helicopter technology to control the twin blades which are placed down-wind of the tower.

More significantly WESCO aims at avoiding the inefficiences of electrical generators (they give about one half of the power it takes to drive them) by generating heat rather than electricity. The rotary motion of their mills will be used to drive friction devices, compressors or heat pumps, alone or in combination.

Granada Television's 'A House For the Future' team had the same idea. Initially their 4½ metre twin-bladed windmill will feed the output from its 24 volt bus alternator into a pair of low-voltage immersion heaters in the domestic hot water tank. A second solution, however, aims at converting the shaft power (about 1hp in a 6 metre per second wind) directly into heat by making it turn the paddles of a water-filled 'churn'. This will heat the water by friction and the hot fluid will be circulated through a heat exchanger in a water storage tank. Though the device still has to be tested it should have the double advantage of an output proportional to *any* wind speed without the need for the very high revs needed by an alternator.

Against this, it must be said that, for a variety of reasons, none of the big windmills constructed since about 1940 have been able to stand up to freak weather conditions. Every one of them has failed mechanically, usually in winds of gale or tornado force, but sometimes due directly to the strains on the structure caused by high-speed operation. In addition, while the power extractable by a windmill is proportional to the area swept out by the sails, so that a 30-metre diameter mill will generate four times as much power as a 15-metre mill, and so on, the weight, cost and demand for materials rises in proportion to the volume, so that the 30-metre mill takes eight times as much material as the 15-metre model to build.

It seems therefore that the immediate future for wind power is the provision of electricity for single homes or small groups of houses, running all the electrical services except heating. Pages 26-28 give more practical details of such use.

Chapter 4
Water Power

Flowing water is an obvious source of power, and watermills have been in use for grinding corn for centuries, with their large wooden wheels driven by a stream of water in the millrace. Unfortunately, although picturesque, these mills were rather anti-social for other users of the river, because they involved the construction of a dam to hold up the water and direct it to the wheel.

The Pelton wheel. Water at high pressure strikes the cup-shaped buckets and drives the wheel at high speed

The same problem of interference with other people's rights dogs anyone who wants to use water power now. While hydro-electric power is very important for some national schemes, the individual who just happens to have running water near at hand cannot assume that he or she has the right to use it for power, without restriction. You need to check very carefully on the rights of the people upstream and downstream of you before you fit up any power device, and you may find that the River Board are also concerned.

However, if you find that you have a legal right to use the water, and are content to have a fairly small installation, you can get a reasonable amount of useful power from water without having either a raging torrent or a 15-metre waterfall. Modern water turbines are much

The Francis turbine. This water mill is more suitable for low water pressures

more efficient than the old water-wheels, and take far less water: you can buy generators that give anything from 50-1,000 watts, and some of these are as much as 90 per cent efficient in converting the power (see Further Information, Suppliers on page 77).

Of the various types of turbine, two main ones are important. The Pelton wheel is useful if the supply of water has a high head—over 30 metres is best—though the actual rate of flow need not be very great. The Turgo Impulse Turbine is a particularly efficient modification of the Pelton wheel, catching the water in its specially shaped cups and revolving at high speed, which makes it very suitable for running a generator.

The Francis turbine design is better for water at low heads, below 15 metres. It has a runner with a vertical axis and vanes that are impelled by the water, which usually turns at about 50-250 revolutions per minute depending on the head of water and the amount available. For suppliers see page 77.

The hydraulic ram

For many purposes, particularly running a water-powered generator, a fairly small amount of water at a high level will give better results than a lot of water just drifting along down a gradual slope. The relationship between head of water and power output is complicated, but in general the higher the head, the more efficiently power is generated—a turbine that gives 100 watts with a 3-metre head of water will give 500 watts with a 9-metre head.

Obviously you could use pumps to raise water from a river to a high-level tank, for instance, but this would be self-defeating, because you would have to use more power to get the water up than it could generate on its way down again. However, if the water is to be taken from a flowing stream, there is a mechanism called the hydraulic ram which uses the power of the whole stream to lift a fraction of it to a higher level, without any external source of power.

Essentially, a hydraulic ram consists of a pressure vessel supplied with water from the flowing stream. A loaded valve lets most of the water out again, but not as fast as it is coming in, so pressure builds up. Suddenly, when this pressure reaches a certain point, the valve closes altogether and all the pressure is transferred to an air vessel. The air compresses and forces a spurt of water up a pipe to the high-level tank. As soon as this has happened, the pressure in the ram drops, the outlet valve opens again and the process repeats itself, the action being one of steady flow into the ram regularly interrupted by a sudden check and a rush of water up the pipe to the high level. Obviously the higher you try to get this water to go, the more pressure you have to build up in the ram, so the spurts of water are less frequent.

You can calculate roughly how much water can be raised by a hydraulic ram as follows:

1. Find the working head of the stream entering the ram, say h metres, and divide this by the height to which water is to be lifted, H metres.
2. Measure the flow of water into the ram in litres per minute, say Q litres/minute.
3. The supply to the high-level tank will be $\frac{2}{3} \times Q \times \frac{h}{H}$ l/min. For example, if the head of the stream is 2 metres and you want to raise it to 6 metres, $\frac{h}{H} = \frac{1}{3}$. If the flow is 2000 litres per minute, the amount lifted will be $\frac{2}{3} \times 2000 \times \frac{1}{3} = 444$ litres per minute. 1556 litres will be returned to the stream, having given up a little of their energy of flow to raise the rest.

The water so raised can be used to run a turbine, and the outlet to the turbine returned to the stream, thus not making much difference to the total flow. The hydraulic ram needs no external source of power, as all the movements of the valves are achieved by the water pressure of the stream, and it needs very little maintenance. In most waterways it will need to be fitted with a filter to keep out fish, weeds, mud, etc., that could clog the valves, and this filter will need to be cleared regularly. The ram itself should be examined for corrosion or growth of algae, etc. annually or more often if the stream is polluted or slow-moving.

The usefulness of hydraulic rams does not stop at raising water for hydro-electric power, of course. They are very useful for lifting water from a stream or river for irrigation, fish-farming or even such things as water-gardens. In this case, however, where the water is likely to be heavily contaminated before it is returned to the stream, you should check even more carefully on your legal obligations before installing any equipment which interferes with the flow or state of the water.

Chapter 5
The Heat Pump

As with water power, one of the troubles with most heat supplies is that they are so often at too low a level to be of much use—the energy is in the form of what engineers call 'low-grade' heat. A tremendous amount of energy is wasted because it is low-grade heat and no one knows how to put it to a practical use—while it is a few degrees warmer than its surroundings, the difference in temperature is not sufficient to run a heat engine, or to supplement domestic heating, etc.

Think, for example, of the everyday event of emptying a bath or washing machine. The water is obviously warmer than its surroundings and on a cold day you can see it steaming round the drainpipe as proof of this. However, even if there were some easy way of purifying the water and removing soap, etc., it would not be hot enough to be of much use in the central-heating system, and certainly not hot enough to drive a steam engine. Yet that bath of lukewarm water may contain more actual heat energy than the contents of the average domestic boiler at 60°C. This is the case with most heat systems: the waste low-grade heat is usually there in considerable quantities, but the temperature is too low for it to be immediately useful.

Our wastage of heat in the home is only a small fraction of the total amount of energy lost in the form of low-grade heat. All machinery wastes energy in this way, and most of it wastes more than it uses. A good steam engine, for example, may produce in power only about 15 per cent of the energy fed into it in the form of the fuel put into the furnace—the other 85 per cent of the heat is lost in warming the surrounding air and the cooling water. (The early steam engines, like Newcomen's big beam engines, were considered excellent if they converted more than *1 per cent* of the heat input into useful work.) Even modern equipment is a long way from perfect—if you have an electric motor that is delivering 500 watts in the form of useful work, you will find that it is taking in about 800 watts from the mains: the other 300 watts are dissipated as heat, a few degrees above the surrounding temperature.

Because of the scale of their operations, the greatest wasters of heat energy are our electricity generating systems. The Central Electricity Generating Board (CEGB) burns fuel equivalent to about 90 million tonnes of coal per year, which makes it the largest single user of fuel in Europe, yet less than a third of the energy in the fuel arrives in homes, factories or offices as useful electric power. The rest, equivalent to over 60 million tonnes of coal every year, is dissipated in the atmosphere over the huge cooling towers, or via cooling water into the rivers. The Thames below Battersea and Lots Road power stations is a few degrees warmer, and as nothing is done with this low-grade heat it represents an enormous waste of fuel.

There are a number of things that ought to be done to conserve this energy—many power stations in other countries pipe their heated cooling water to districts of houses, sometimes through hundreds of kilometres of pipes, and the CEGB is by no means as efficient as many other generating bodies—but the fact remains that there will always, with such a large mass of energy-consuming machinery, be a certain amount of waste heat left over at temperatures that are useless for the conventional heating system or engine.

The heat pump is a device that can take in such low-grade heat and concentrate' it into a smaller mass of material at a far higher temperature. This means that heat pumps are peculiarly useful in a densely populated and industrialised country like Britain, where a lot of low-grade heat is constantly issuing from houses, factories, power stations, vehicles, etc.

Heat pumps are really refrigerators run in reverse. In the ordinary refrigerator we put food at room temperature into the cabinet, and the cooling coils (the 'expansion' stage, see page 23) extract heat from it until it is near freezing point. The extracted heat is disposed of by circulating it round to the back of the refrigerator, where another set of coils or pipes (the 'compression' stage) is exposed to the air: these pipes dissipate the heat into the kitchen. If you have a large refrigerator or deep freeze in a fairly small room you must have noticed how much heat is generated when the room is shut up overnight. It is essential to allow the heat to escape easily, otherwise the cooling stage will not be able to function efficiently.

In the heat pump, the source of low-grade heat (which may be air, water or the ground next to the building) is cooled in the expansion stage by passing the air or water over coils or burying coils in the soil. The heat extracted is passed round to a compression stage inside the building. For example, if the compression coils are in a house, they can be fitted with a fan and air ducts to circulate warm air over the whole building: in a factory, the coils could form part of a heat-exchanger to start heating water or other materials that need to be heated.

Heat pumps have one great advantage over refrigerators in terms of efficiency. A refrigerator has to have a pump or heater to compress the fluid in the compression stage and waste heat will be generated by these mechanisms. This has to be disposed of as well as the heat extracted from the food, and if it is not, the refrigerator will not run well. In a heat pump, on the other hand, we actually want the heat produced in the compression stage, and any waste heat from the mechanism will simply be added to the rest and made use of.

Almost any material can be made to give up its heat to make a source for a heat pump. Obviously it is better if the source starts off reasonably warm, although systems have been known to get useful heat out of air at -20°C. Therefore it pays to prospect carefully to find out whether there are any outlets for waste heat in your area, e.g., streams or rivers that have been used as cooling water for factories and therefore slightly warmed themselves, currents of warm air from other buildings, or even just places where the air or water are warmed by the sun more than the average. It is well known that the temperature a few metres below the surface never gets as low as at ground level (that is why water pipes are sunk in the ground, to avoid frost) and coils buried in the soil or down a well can collect heat better than at the surface. On the other hand, earth as a source of heat has the disadvantage that once you have extracted as much heat as possible from a given mass of soil, it may take a long time to warm up again, while air and water are constantly being replaced by fresh supplies. The main sources for heat pumps may be summarised as follows:

Air

Freely available, usually warm enough to give efficient operation, causes very few problems in the collecting coils. *Disadvantage:* the collecting equipment has to be bulky, with large fans or air-pumps (because even a small amount of energy heats a large volume of air).

River water

Only available in selected places, but may be warmed by industrial activity upstream, giving a free source of extra heat. Constantly changing and therefore never likely to become cooled to a point where heat extraction is inefficient. *Disadvantages:* may be polluted, muddy or corrosive and so cause various problems in the collecting system. Cannot be cooled below freezing point, which could easily happen if the system is large and the river near this temperature when it comes into the system, though thermostatically controlled switches could easily solve this problem.

Well water

More readily available than river water, can be provided by sinking a well, although this is an expensive capital undertaking. Water is usually considerably above freezing point even in winter. *Disadvantages:* usually even more likely to cause scale or algae than river water. Can be frozen rather easily, and there is no quick change of water.

Earth

Readily available by sinking coils in the ground. Usually above freezing point even in winter and not necessarily stopped from supplying heat even if lightly frozen. *Disadvantages:* source is limited as a large volume of soil gets cooled and takes time to heat up again. Corrosion may take place unless collector coils are carefully designed.

Solar panels

Usually a source of heat well above the average in temperature, clean and non-corrosive. *Disadvantage:* no supply of heat at night and little on very cloudy or winter days.

Of all the systems, the last (solar panels in conjunction with a heat pump) seems the most promising, but the others all have their advantages, especially if there is a local source of low-grade heat.

It is normal to measure the efficiency of heat pumps by comparing their output with the amount of energy necessary to keep their compressors and ancillary equipment running. Under unfavourable conditions, for example in the winter months and in a situation where there is no local source of waste heat to be drawn upon, the efficiency is usually about 1.7-2.0, meaning that for every 1 kilowatt of electric or other power you put in, you can expect 1.7-2.0 kilowatts of heat power out. Obviously this still represents a useful saving of energy, but as the capital cost of the heat pumps is at the moment a lot more than that of conventional heaters, not many people would want to install them if that efficiency was the limit. Fortunately, where conditions are better, the efficiency may rise to 6 or even 8, so that you get out much more heat power than you put in.

One of the arguments against the use of heat pumps on a wider scale is that they depend on heat which is wasted from industrial and other processes. Therefore, the argument runs, if these processes are eventually made to run more efficiently, there will be no waste heat for the heat pumps to utilize. There are two counters to this argument.

First, the business of making our industrial processes more efficient, in particular our electricity generating techniques, seems likely to take at least twenty years, and probably longer. Heat will go on being wasted for all these years unless we take steps to rescue some of it by installing heat pumps.

Second, heat pumps, when developed to be highly efficient and not too bulky, would be the ideal way to make the low-temperature heat from solar panels useful for a whole range of devices, both for home and industry. For example, at present the heat from solar panels is not usually at anything like the temperature necessary for cooking or baking bread, so even enthusiasts for solar energy have to use coal, gas or electricity to run their stoves and ovens. By concentrating the heat of the sun, using a heat pump and one of the modern heat-transfer media that can circulate at several hundred degrees without boiling, it is possible to take the heat from a panel at, say, 60°C and raise the temperature to about 171°C, a good baking temperature. The equipment is not yet commercially available.

While commercial heat pumps are rather costly at the moment—being anything from eight to ten times the cost of a conventional heating plant with the same output—this is largely caused by the much smaller demand for heat pumps, which tend to be made on a 'one-off' basis. Harold Gee, one of the enthusiasts who have helped to build up the whole movement for alternative ways of handling energy, has constructed a highly efficient heat pump for his home, using secondhand refrigerator and heating parts. He works with air as his source for most of the year, blowing it with fans through a large collector like a big car radiator, but when the air temperature drops in winter he uses two 'wells' which he dug out and lined with concrete himself. Water is extracted from the first well, cooled by the heat pump, is then returned to the second well and allowed to flow naturally through pipes in the soil, picking up heat as it goes. It ends up warmed in the first well.

He claims that this plant gives an efficiency of about 5.0 - 5.5 kilowatts of heat power out for every kilowatt of electricity used to run it.

Chapter 6
Methane Generators

When any kind of vegetation decomposes, bacteria are breaking it down to simple substances that can be used again by future generations of plants. Most of the time the main product is carbon dioxide, which disappears into the atmosphere, and a little ammonia, which is rapidly taken up by the soil or by other bacteria. However, this type of reaction needs oxygen from the air to keep it going, and sometimes the vegetation is cut off from the air—it may be buried, or under water in a river or flooded area, or sunk in a marshy spot. In such cases the bacteria that develop produce some carbon dioxide, but more of the inflammable gas methane.

Methane is sometimes called 'marsh gas', because plants buried in the mud of marshy spots produce bubbles of the gas that sometimes 'plop' to the surface. If lightning or some other cause sets them on fire, the flame will sometimes run for hours from one bubble to another, making a ghostly flickering light over the marsh, called the will-o'-the-wisp. In some places large deposits of prehistoric plants have been buried deep by earth movements and these produce subterranean caverns full of methane gas under pressure, for example North Sea gas.

The gas can develop quite quickly if conditions are right. Those responsible for rubbish tips know that methane can be generated in large quantities deep down in the pile, and there have occasionally been quite serious explosions when workmen have uncovered a giant 'bubble' in municipal refuse tips. Sewage can also produce large quantities of methane if it is not sufficiently aerated during the purification process, and again the experts on sewage disposal have learned to regard the gas as a hazard.

In fact, methane generation is really an example of the way in which the sun's energy is available to us in all sorts of ways, if only we learn to use it properly. The sun warms the dying vegetation, and this warmth encourages bacteria to grow, and, as they grow, to convert the almost unusable waste into a highly inflammable and effective fuel. As natural gas users know, the actual heating power of methane is greater, per cubic metre, than the older 'town gas' made from coal. Even the mixture of methane and carbon dioxide that is the first product of the bacterial action (and is often called 'bio-gas' to distinguish it from pure methane) is a better fuel than town gas, and many people and organizations are now making use of bio-gas as an important by-product of waste disposal.

Bio-gas as produced from most wastes consists of between 64 and 72 per cent methane, 24 to 32 per cent of carbon dioxide, and traces of hydrogen, nitrogen and other gases. Sometimes it contains sulphur compounds that make it smell unpleasant (particularly if the wastes contain animal matter such as offal), but fortunately this does not happen with most samples of bio-gas. To make it, you need primarily a generous supply of waste organic materials such as vegetable trimmings, grass cuttings, straw, wood shavings, and even processed materials such as paper and string or rags (cotton or wool)—in fact, almost everything except man-made fibres and plastics. This vegetable waste is best mixed with sewage of some kind—cow or pig dung, chicken droppings or domestic sewage. This not only increases the amount of organic material available to be fermented, but also provides large numbers of bacteria to start the process. In the absence of sewage, the right bacteria will eventually find their way into the vegetation from air and water, but this takes longer.

The waste is churned up with water into a slurry, to assist the work of the bacteria, and placed in a container called the *digester,* which is covered tightly to keep air out and bio-gas in. A digester can be as simple as a hole in the ground with a domed top floating on the slurry, but such an arrangement is difficult to clear out for the next batch of vegetation, and in this country it is better to have some means of applying gentle heat to speed up the bacterial action. Methane production is at its best when the mixture is at about 32°C, and although the mixture gets warm by the fermentation, an outer jacket that can be filled with warm water will improve the efficiency greatly.

A stirrer or agitator in the digester keeps the production of gas even: this need be no more than a kind of paddle working through a hole in the lid of the digester, as long as it is a tight fit to prevent gas leakage. Small plants such as the Bio-Gas equipment (see page 77) are often made out of plastic containers of various sizes, a medium-sized one for the digester in a larger container that acts as a warm-water jacket, with a third container inverted over the top to trap the bio-gas. Extraction of all the available gas takes from one to four weeks, depending on the temperature, the type of waste, and the amount of sewage added to give an initial charge of bacteria.

The gas is led off from the upper container through a tap and plastic pipe. If you have a large digester it is best to run the gas into a plastic storage bag made of butyl rubber (these are available from the manufacturers of the digestion equipment), and distribute it from this. Alternatively, a series of smaller units can be set up at weekly intervals so that there is always one unit in full gas production, and the gas can be led directly to a burner (for instance, a greenhouse heater).

Methane is not liquefied by pressure (unlike butane, for instance), but it can be compressed into fairly light-gauge cylinders for transport. Again, the suppliers of large-scale equipment also supply suitable compressors.

One advantage of methane production is that sewage treated in this way gives a sludge that is still useful as a fertilizer, which has lost almost all its smell. The Greater London Council, for example, has a plant that can produce about 164,000 cubic metres of bio-gas every day from sewage, with a fuel value equal to about 113,650 litres of petrol, or over 40 million litres per annum. After the process, the sludge left is almost odourless, but contains 9 per cent of nitrogen, 2½ per cent phosphate, and 1 per cent potash, so that it is quite an effective fertilizer.

The figure given above for the 'petrol equivalent' of methane underlines one of its most valuable properties as an alternative source of energy. Unlike solar and wind power, which are essentially forms of energy for static situations, methane can be used to run a conventional vehicle engine, with a modification of the carburettor. The compressed gas, in cylinders, can be carried in the boot of a car, or a trailer—some taxis are equipped to run on methane or diesel fuel, the diesel being used only for emergencies if methane runs low a long way from base. Methane has, weight for weight, rather better fuel performance than ordinary petrol, and even the impure bio-gas is not far inferior.

Chicken-farmer Harold Bate originally ran his Hillman car on methane produced by his farm wastes and claims that anyone could do the same with a very simple modification to the conventional carburettor —he has in fact developed a second line of business from his Totnes farm, selling the modified jets, etc. Professor P. D. Dunn of Reading University has pointed out that if other livestock farmers followed this example, the methane from Britain's cow-dung alone could replace half the imported oil used by the agricultural industry (not only for running cars and tractors, but for heating animal houses, batteries, etc.). This would be equivalent to about 2.5 million tonnes of oil products every year, and the sludge left from the process would still be suitable for adding to the fertility of the fields. Such a process would make a lot more sense than the present practice of many mixed farmers, who pay someone to take away and tip all the farmyard slurry and then pay large sums for artificial fertilizers to put on the land—fertilizers that themselves consume a great deal of energy in their manufacture.

On the other hand, for the ordinary householder, there are some important points to be made against methane generators on the domestic scale. First, they need quite a lot of material to keep them going—excellent if you are a farmer or market gardener with a mass of waste products to dispose of, but not so easy to find for the average family. Each kilogram of *dry* waste will give about 0.36-0.50 cubic metres of bio-gas, but you have to remember that most vegetation contains about four-fifths of its weight as water, so you need 2¼ kilograms of undried waste for that quantity of gas. If you use your own domestic sewage (a messy business, and not one that would meet with much approval from the health authorities), you will not gain much: a simple rule is that the sewage produced by one person in a day will make about 0.028 cubic metre of bio-gas, so it is hardly worth the health risk. Apart from that, if anyone in the house is taking large quantities of antibiotics, these may affect the methane plant and kill the bacteria, so that the plant has to be cleared out and the batch started again.

The second problem is keeping a small amount of waste at the right temperature for methane production. On a large scale the natural heat of fermentation builds up fairly well, but in a small domestic plant you need hot-water jackets and similar aids, which means expending more fuel and effort.

Third, and most important, methane catches fire easily and is explosive. Before you embark on setting up a plant, consider carefully whether you will be able to run it with all the necessary precautions. Think of the matter from the point of view of your neighbours—not many people relish living next door to a gasworks, and how do you feel about living next to an *amateur* gasworks?

Methane production is probably best suited to a community or fairly large farm with land that can provide plenty of vegetable or animal waste all the year round. The methane can be used for heating a greenhouse, using a conventional gas-heating plant (although you may have to make sure that your gas does not contain too much hydrogen sulphide and other sulphur compounds, otherwise the plants could suffer). However, you must also consider the fire and explosion hazards: methane is exactly the same as North Sea gas, and requires exactly the same precautions. Probably if your plant is in the open air it cannot cause too much trouble, but you must take proper action to prevent a flame blowback from your heater to the digester. It is perhaps significant that Bio-Gas Plant (see page 77), suppliers of equipment for experimental or small-scale methane production, are careful to specify that they accept no liability for injury, loss or damage arising directly or indirectly from the use of their products.

Chapter 7
Recycling Waste

Another solution to the problem of recycling sewage, apart from methane production, is to convert it to usable compost in the house. This may sound like a return to earth closets and the visits of the euphemistically named 'nightsoil collector', but modern equipment has made it possible to process sewage to a virtually odourless compost, suitable for use in fertilizing the soil, without sacrificing any of the comforts or health benefits of the conventional lavatory system.

The problem, for society, is that we use so many resources to get rid of our sewage. The flushing toilet uses, on average, about 59,097 litres of purified water per year, and the actual material that is discharged, although it has a high content of nitrogen, phosphates and other useful nutrients for plant growth, is often entirely wasted by being discharged into the sea.

Two alternative systems, the Clivus Multrum and the Mullbank self-contained toilet, are available (for instance, from Conservation Tools and Technology, see page 77), and both use a system whereby bacteria in an aerated composting chamber break down the sewage (and also any kitchen waste) into an odourless compost. A fan system keeps the bacteria correctly aerated, which ensures that methane production cannot take place (with a consequent risk of explosion), and the air is vented to the outside of the building so that no smells can escape into the bathroom or the house generally. The compost is cleared out about every twelve months, depending on the rate of use of the equipment, and is immediately suitable for fertilizer for the garden or agriculture.

While it appears that health authorities are reasonably wary of the idea of installing such equipment in the average house, and virtually cutting it off from the public sewer system, there are many advantages in country districts where main drainage is not yet available. The cost of one of these devices is less than that of having a septic tank built, partly because the self-contained units need very little in the way of such things as service pipes, and they are considerably more hygienic than chemical closets or earth closets.

Chapter 8
Recycling Water

Most of us think of water as something that comes out of a tap, instantly available, clean, wholesome, and in unlimited quantities. The true facts are that every litre of water we so lightly squander has been collected and purified by a complex organization that is approaching the limits of its capacity to keep pace with our demands for water. More and more parts of the country are having to accept some degree of water rationing during the summer months, and some are becoming marginally short of water all the time. If we go on insisting that we must use over 159 litres of water a day for every man, woman and child, these shortages are soon going to become country-wide.

One of the troubles is that we use the same quality of water for everything. Our taps are supplied with water fit to drink—filtered, purified, chlorinated to kill bacteria and perfectly safe for food preparation. However, we use this pure water not only for these purposes, but for washing the car, watering the garden and even mixing cement and a host of jobs for which a 'second-grade' water would be quite adequate.

While drinking water is still a desirable service for the house of the future, we should not have to demand 159 litres a day. Water for

Collecting water on a farm in the Queensland outback. Every drop is conserved and filtered in the butts

the secondary purposes could be recovered and used again, and rain or surface water collected.

Some jobs would only require properly gathered rain water. Most of us waste the water that falls from the sky—if we have a butt, it is usually of about 181 litres capacity at the most and soon overflows in the rainy seasons. A large tank collecting all the roof drainage could be used for garden water and odd jobs for most of the year, like washing the car and mixing concrete, without ever turning on the main tap.

For other purposes, some degree of purification is necessary, and the best means on the small scale is distillation, using the sun or some cheap fuel as the heat source. The system is simple: the water which is to be purified is heated until it begins to evaporate and the vapour rises, leaving behind any impurities such as dirt, salt, detergents, bacteria. The vapour is allowed to contact a cold sloping surface which it runs down into a receiver, which in a small still may be a bottle or, in a large one, a system of gutters leading to a reservoir. On the island of Guernsey, chronically short of fresh water, about 2.25 million litres per day of distilled water can be added to the island's supply from a plant that purifies sea water, using the cheapest fuel oil as heat source.

From the ecological point of view, the best heat source is the sun, and a great many solar stills have been designed and operated. A still producing 2,727 litres of water a day was set up in South America over a hundred years ago; brackish water ran through shallow troughs under a system of sloping glass plates like garden cloches. When the sun struck the system, pure water vapour condensed on the glass plates, from which it ran off to another system of troughs at the side and thence to a storage tank.

A popular design for solar stills follows the same pattern, but uses a dome or tent of plastic as the condensing surface. Water in a pool below the plastic is evaporated by the sun and condenses on the inside of the plastic. It runs down into a ring-shaped trough under the edges of the plastic and then into a collector. Such a system is used in the air-sea rescue solar still which forms part of the survival kit for shipwrecked people: it will take in sea water and produce enough pure water every day to keep two or three persons alive.

A combination of this system and the South American design can be adapted for use in any garden. Instead of salt or brackish water, waste water from the house is allowed to trickle down a long trough, which goes to and fro (always gradually downhill) until it comes to the ordinary drain or soakaway. The trough is covered by a kind of long cloche made of plastic stretched over wire hoops, and the edges of the plastic are arranged so that drips of water fall into two narrower troughs, one at each side. These eventually lead to containers for the fresh water or to a permanent tank sunk in the ground. As the sun warms the system, water evaporates from the central trough and condenses on the plastic, and from there runs down into the gutters at the side. The water should be clean enough for almost any purposes, but it can, of course, get contaminated by bacteria from the air and could also be mixed with rainwater falling on the outside of the cloche. Hence, it is probably wise not to use it for drinking except in emergencies, and then it should be boiled.

As it is not important that all the water is evaporated, the rate of flow of waste water should be arranged so that some water emerges at the other end of the system and runs to waste. This will mean that there is a constant flow throughout the central trough, and debris or residues from the water will not build up very fast: even so, with a few metres of such a system, quite an amount of water can be recycled.

This simple still can be constructed of plastic guttering for the troughs (if you use the black variety it will absorb the sun better, for evaporation). The plastic over the top can be almost any kind of plastic film, such as polythene, but there are on the market one or two varieties with a slightly 'textured' finish that helps the condensed water to run easily down into the collecting troughs. This not only ensures that the water is collected smoothly from the surfaces, but also that the plastic does not become too misty, which could interfere with the absorption of the solar heat.

Air-sea rescue solar stills being tested by the Royal Naval Survival Equipment School

Chapter 9
What can you do Now?

After looking at all the fascinating 'hardware' that Brian Trueman demonstrated in the Granada series, all aids to an alternative style of living, you may well be wondering why these ideas are not already being used in more houses, or at least being installed with all urgency. After all, solar heating, wind power, water power, and so on, are not new ideas, and even the equipment for making use of them has not changed much except in the choice of materials—plastic instead of metals for solar panels, GPR (glass fibre resin) instead of pitchpine for windmill sails. Possibly the most 'novel' invention that has potential for self-sufficiency, the solar cell, is at the moment the least important as a practical source of energy.

So why the present crisis in energy supplies?

One of the answers is, of course, apathy. We have had our fuel so cheaply for years that we have not bothered to take precautions to save it. Electric heating and electric air-conditioning in buildings that have only rudimentary insulation is one of our greatest extravagances, but there are countless others—picture windows that act as vast radiators of heat to the general neighbourhood, for example. Now that fuel has gone up we feel rather like a man who has been running a Rolls Royce for years at the company's expense, and now suddenly finds that he has to buy his own petrol.

But the main reason for the slowness in practical self-sufficient development is that the problems connected with the equipment are very different from those with conventional heating and power plants. Solar and wind power are variable in ways that coal and oil are not, and we have to find strategies to deal with this variability. In fact, the problems of self-sufficiency make us realize that one of the greatest advantages of coal, oil or nuclear fuels is that they are very efficient ways of *storing* energy for future use. A pile of coal in the backyard is a far more reassuring guarantee that we shall be able to keep warm next week than a weather forecast telling us that the sun may possibly be strong enough to warm up our solar panels then. Storage of the alternative sources of power is in fact the biggest problem to be solved, and it is fair to say that the solutions proposed so far are not as effective as our conventional coal-stacks or oil-tanks as reliable energy 'banks'.

Even so, there is a tendency for writers to exaggerate the difficulties of the new 'soft' technology just because it is new. I have seen it written at least a dozen times that wind-powered electricity cannot work because of the variability of the wind, and the fact that storage batteries to even out the supply would be 'prohibitively' large, heavy and expensive. The facts are that about 50 kilowatt-hours capacity, enough to keep the average household going during very calm periods when the wind contributes nothing, can be stored in batteries that occupy about 0.8 cubic metres, weigh about 2.5 tonnes, and cost around £1,500, with a life expectation of 10-30 years, according to the type of battery. All of these figures seem large at first, but whether they are prohibitively large depends on one's standards. Apart from the weight, the same sort of space and cost arrangements would have to be made for a second car in the family, and this contributes nothing to energy-saving, while the batteries could well be part of a system saving virtually all the conventional electricity costs.

Similarly there is a tendency to consider that the storage of heat, as collected by solar panels during the warm seasons, is quite impractical because of the 'enormous' amounts of water or other storage material involved. The actual figures, although certainly large, are not terrifying when you consider that such a scheme could make a house entirely independent of conventional fuels. Taking the quantity of water calculated by Alex Pike and his colleagues in the Cambridge University Department of Architecture, about 45,000 litres of water should be enough to give effective storage of solar heat from warm periods to cold ones, and thus level out the peaks and troughs of solar heating. This means $45m^3$ of storage, no bigger than a single garage. The storage container must have as small a surface area as possible—a sphere is ideal.

If the hopes of enthusiasts for latent heat or chemical storage come to fruition, the storage tank could be between $^1/_3$ and $^1/_5$ of the volume. This system, which is at the moment entirely experimental, depends on the fact that many materials take in large quantities of heat as they melt and radiate it outwards again as they resolidify on cooling. Thus if a hollow wall is filled with such material, and heated up when a solar panel is working well, the wall will continue to radiate large amounts of heat when the solar supply has been cut off. The greenhouse on Granada's House for the Future will have a chemical storage unit to help maintain an even temperature at night.

However, although the case for the relatively self-sufficient house is a very strong one, most people are going to have to reach their 'house of the future' one step at a time. The following sections deal with the series of moves that we can all make in the direction of self-sufficiency, gradually getting less extravagant with energy, making ourselves independent of power cuts and fuel shortages and at the same time adding to the comfort and general efficiency of our homes.

Putting up the greenhouse supports for the Macclesfield house

Chapter 10
Planning—Who needs to know?

Before you embark on equipping your house with any of the alternative energy sources, or start to reconstruct it to make it more convenient to live in and less extravagant on fuel, you will have to plan the operation very carefully. Most of this planning is for your own benefit: you need to decide precisely what you expect the changes to do. Are you installing a solar panel just to save a little money, for example, or do you expect it to make you free of the costs of oil or coal altogether? Is your wind-generator there as a source of emergency power when there are power cuts or shortages, or do you expect it to provide you with enough electricity for all your needs?

Apart from the capital costs that you may have to find, and the amount of trouble and disruption in the house while you are doing the work, there are plans that need to be made for running the new equipment, because the problems of alternative power sources are different from those of conventional energy. A large windmill, for example, needs a certain amount of watching, particularly if the weather is squally, and if you and your family are out of the house a good deal of the time you may have to immobilise the sails on some days to make sure that the mill does not over-run. This will mean that the overall yield of power will not be as great as if you were able to run it almost all the time.

One of the reasons why alternative energy sources such as solar roofs and windmills usually run fairly well when set up by communes is that life in a commune means that there is nearly always someone to watch the equipment, and see that it is running properly at all times. This becomes increasingly important if the equipment is rather experimental and could do disconcerting things. Life in a large group such a commune also lends itself to greater flexibility than the average family can manage. For instance, the group known as BRAD put up a large solar roof (designed by Robert Vale). During summer afternoons, they found that it collected so much heat that everyone who was available took turns to have baths just to use up the hot water. This would hardly work if most of your family are out from nine to five, and you would have to include some kind of 'dump tank' in your design to get rid of surplus hot water, or have automatic shutters that can be drawn over the solar panel surface when the water gets to a certain temperature. The Macclesfield house has ventilators in the ridge of the roof to deal with excessive summer heat, radiated into the house by the collector panels when the water system is turned off once enough hot water has been collected.

When planning any equipment that will take energy from your surroundings, such as a heat-pump using a river or stream as a source of low-grade heat, or a stream operating a hydraulic ram or water turbine, you must also work out carefully what effect this will have afterwards. Cooling a stream may make it freeze more easily, slowing it down could make it silt up—both will affect plant and fish life, and the amenities of people downstream.

Windmills and similar structures could be dangerous if they are not very carefully designed and anchored to the ground so as to withstand the strongest wind. Your neighbours, if they live close to you, may have very reasonable misgivings about a large windmill spinning away where any accident could send a chunk of wood or fibreglass flying through their greenhouse or over the heads of their children, so be prepared to explain exactly what you are building, how it works, and what precautions you are taking to keep it stable and in one piece.

If you make sure that people understand what you are trying to do, you may even find some converts who also want to play their part in saving energy, and start to establish an alternative technology group in your neighbourhood. In general consider and respect your neighbour's rights. Don't, for example, put up any building that will keep the sun off his house or garden.

Planning authorities

Apart from such fairly obvious reasons for consultation, there are a number of projects that will need discussion with your local Planning Authority. Basically they will want to know about any work that you intend to do that alters the construction of your house in any way, or changes the external appearance of the building. For example, if you wanted to build a tower on your roof to mount a windmill, you would definitely need permission from the local authority, and before this was granted you would have to satisfy them that the structure would not be an eyesore, would not put too much strain on the timbers in the roof, and was not likely to cause an accident. Similarly, if you want to alter your roof to make a solar collector, or just mount some solar panels on it, this would change the construction and the appearance of the roof, and you would need permission.

Constructions that are not 'buildings' and are separated from the house do not normally require permission. For example, you could put your windmill on a tower in the garden, or erect your solar panels behind the house and connect them to the hot water system by lagged pipes.

You can also put up storage buildings without permission, so if you needed a small place to store electric batteries, for example, this needs no consultation. There are one or two regulations to be followed in the siting of such outside structures: do not put anything high where it could cut off the light from windows in the house (not nearer than 3.6 metres) or any building less than 2 metres from the house walls, but otherwise 'detached' equipment introduces very few planning complications.

For those structures that do need permission, the best thing to do is to go and discuss your plans with the local Planning Officer before you go very far. He will want to know what the installation looks like, how it is to be constructed, and what precautions you are going to take to make sure that it cannot spoil the amenities of the house or neighbourhood. Take with you the following information.

1. A drawing of the proposed construction. This does not have to be a beautifully-produced architect's plan, as long as it is to scale, clear, and shows constructional details. For instance, if your installation is a solar roof, have a scale plan showing how it relates to the rest of the house, and a section showing the materials to be used, their thickness, how, for example, the collector will be mounted and braced to the roof timbers, and how you intend to waterproof any joints between the collector and the rest of the roof. You will need all these details for your full set of plans, so it does no harm to work them out fully beforehand, and you may well realise something that you have forgotten to take into account, while you are drawing up even rough scale plans.
2. If you are using a commercial product, such as a ready-made solar panel or windmill, take along a photograph or manufacturers' brochure to show exactly what the installation will look like. Most of the manufacturers who are operating in the field of alternative sources of energy know that planning officers and similar people need to be convinced of the practicality of their equipment, and they will often not only supply a photograph, but tell you where you and the planning officer can see their equipment in use.
3. Although the efficiency of your installation is not so important to the Planning Officer as its safety and stability as a building, be prepared to discuss the sort of output you expect to obtain from your equipment. In time, it is to be hoped that Planning Officers will get used to applications for the new equipment, and know about the efficiency and so on for themselves, but at present any effort to bring such facts to their notice will be a help to everybody.
4. For the same reasons, collect as much information as you can on alternative technology, to show that your plans are not just a product of your own thinking, but part of a world-wide movement for more rational use of energy.

If you engage an architect to produce your designs, he will of course look after all the contacts with the planning authorities. You may find, if the construction is a large-scale one, that you will have to have several meetings between yourselves, the architect, and representatives of the local authority to determine exactly what you want to do, and how far they are prepared to let you do it.

Planning applications

When you are ready to make a formal application, draw up your plans completely, taking into account any suggestions or modifications that have arisen during discussions with the Planning Officer, and submit these, together with a set of completed Planning Application forms, for the next meeting of the Planning Committee for your area. You will normally need a drawing showing the position of your house relative to others in the road or area (scale usually 1:2500) and detailed drawings showing your house and where the new construction fits in (scale 1:100). If there are any details of the method of construction, or the exact materials to be used for a small section of the work, that you feel cannot be shown with sufficient clarity on the 1:100 drawings, you can either write down the description on the appropriate drawing, or, preferably, make another drawing on a larger scale to show up the particular point. If the Planning Officer has made comments on some such point, for example advising you to use a non-conducting pad between two unlike metals so as to avoid electrolytic corrosion, make sure that your drawings show clearly that you have taken the point. While he advises the various Planning Committees, the Planning Officer cannot be expected to remember details of individual plans when he sees so many in the course of a week.

Unless you are employing an architect, make sure you first read the Building Regulations 1972, and any subsequent amendments, very thoroughly, and note down every regulation that seems to apply to your project (do-it-yourself magazines can be very helpful in providing articles that interpret some of the more obscure regulations). If it seems essential for your project to build something that does not comply with the regulations, make sure that you note this down carefully in your planning application, as you will have to get a 'waiver' of that particular requirement. When you come to start work, after getting your permission, you will meet the Building Inspector who will want to see that you are carefully following the regulations. He can legally force you to pull down your work if it does not comply with the regulations, even by a trivial amount. But if you have studied the Building Regulations thoroughly enough not to make any obvious mistakes, you will get co-operation and advice from the Building Inspector.

You *ought* to get permission or refusal in about two months, but in practice most committees take a good deal longer than this, and there is very little that can be done about it. A lot will depend on the size of your project and how open-minded your local Planning Committee members are.

If you should get a refusal, you can appeal to the Department of the Environment, a fairly simple proceeding which involves filling up two copies of the Appeal form and sending it to the Department with copies of your plans and all the correspondence with your local authority. The Department will appoint an Inspector to look into the case and recommend whether you should be allowed to go ahead after all. He may well want to come and see the property, not to discuss things with you, but to get a better idea of the situation. If he also rejects your application, there is nothing for it but to revise your plans (you should have a good idea by this time of what the authorities object to), prepare fresh drawings, and start again with another planning application.

If this sounds a little depressing, remember that it is the worst sequence of events. If you plan your installation properly from the first there should be no need to go through the dispiriting process of rejection and appeal. Find out as much as possible about the work you want to carry out, from books and manufacturers' literature, and if possible try to find other people who have carried out similar jobs and are living with the results of their work.

Building Societies

If you are buying your house on a mortgage, the Building Society will want to know if you plan any extensive alterations to it. Their interest is rather wider than that of the local authority, because they want to be sure not only that your operations cannot cause the house to fall down, but also that the place will still be saleable when you have finished your work. If you make it look too peculiar, or match it so closely to your family life-style that it is completely unsuitable for anyone else, potential buyers may be frightened off, and this worries the Building Society, who regard your house as the security for their loan to you.

We asked a number of Building Society local managers how they would react if a client walked in with plans for solar collectors, wind-generators, methane plants, and so on, for a mortgaged house. The reactions were very variable. However, all the managers agreed that in any case they would put up any questions of this kind to their head offices, and the general consensus of opinion from the Building Societies was that they would have no objections in principle to any of the ideas, and some would even consider lending extra money to assist with the installation. However, they would need to know the following details:

1. Detailed plans, as described for the planning application, showing exactly what was to be done and what the final appearance would be.
2. Preferably photographs from manufacturers, or from similar installations in other houses.
3. Details of any follow-up alterations in the facilities to the house. This is an important point for the societies, who want to be sure that the house will not be made so unconventional that nobody would want to buy it. Thus (an example quoted to us) they would look with favour on any equipment that could provide an *alternative* source of electricity, but if you decided that your faith in the system was such that you had the company mains cut off, this would not be so popular with the Building Society. Similarly you are likely to get more support for your solar panel or roof if you agree to maintain a conventional heating system as well.

In general the Building Society attitudes we encountered might be listed as follows:

Solar panels

A good idea if they are backed up by conventional heating. Commercial panels preferred to any home-made system, because of appearance rather than efficiency. Some societies might advance money for installation, and they would certainly regard panels in an existing house as tending to increase the value.

Solar roof

Slightly less favourable: probably good if fitted from the outset on a new building, but adapting an existing building might spoil its appearance and lower the value.

Wind generator

Good if considered as an emergency supply of electricity, but considerable doubts about the aesthetic and safety aspects of windmills in the average garden.

Methane plants

Very doubtful because of the safety and health aspects.

Heat pumps

A good idea if they can be made to work for the average house. The societies would need a specialist report before advancing any money for an installation, but would have little objection to the system being fitted at the householder's own expense.

Water power

A good idea in the very small number of properties where it can be used. Could increase the value of the house.

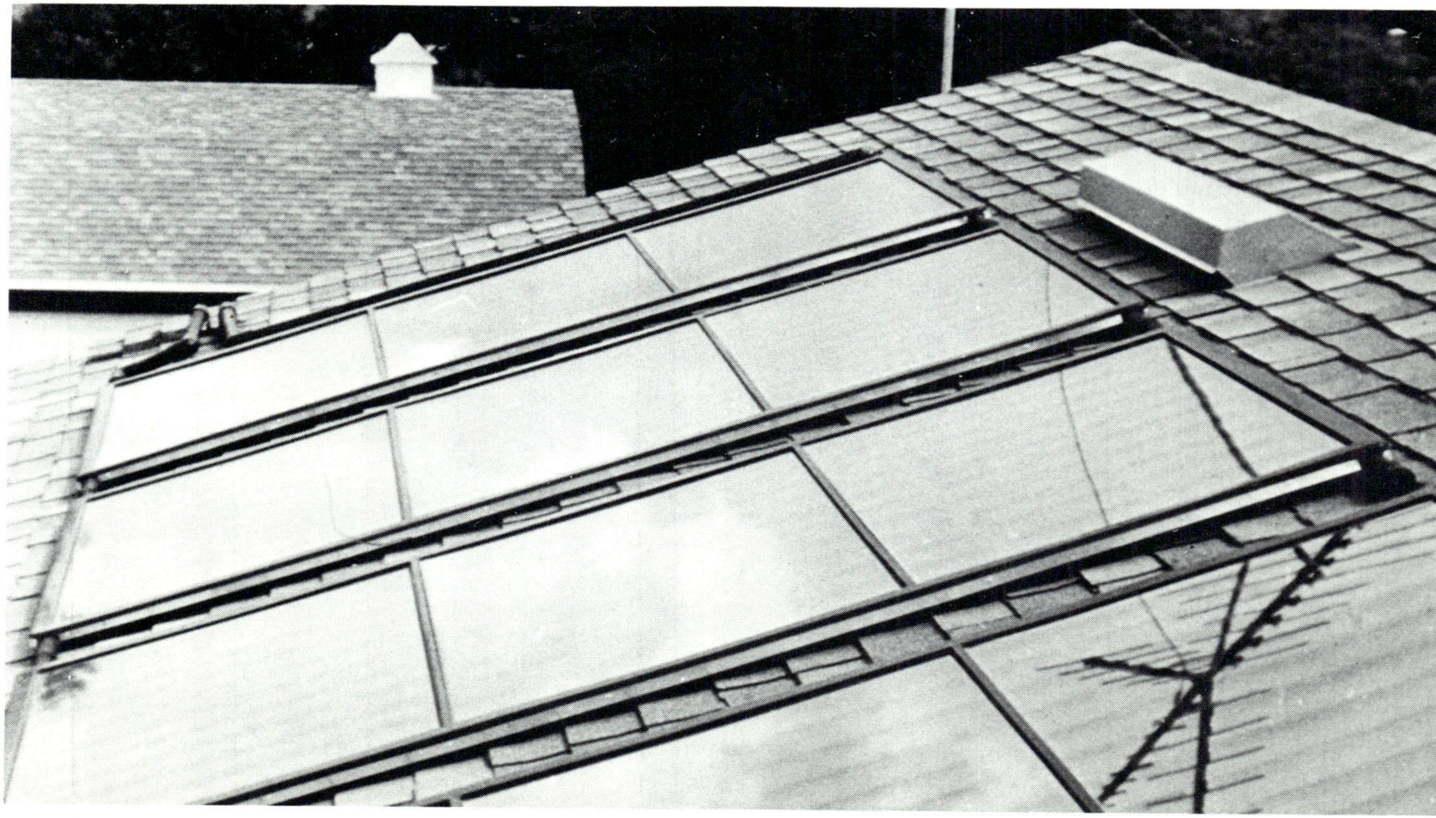

Solar panels on a roof in Virginia heat water for the swimming pool

Chapter 11
Insulation

One of the most primitive features of modern houses is the extremely low standard of insulation, and this is the primary cause of most of our present energy crisis. We pump energy into our buildings, but they are so constructed that it is rather like pouring water into a sieve. Perhaps if heat were visible, like smoke, and we could see it escaping from windows, walls, roofs and doors, we should pay more attention to insulation, but as it is we are only conscious of the losses when the bills come in.

There is in fact a system, known as Thermovision, that makes heat visible, at least on the TV screen. The Thermovision camera does not detect ordinary light, but it can show up the infra-red heat rays as a range of colours, white being hot and blue cold. A glance at some quite modern-looking houses through the eye of Thermovision shows up how far they fall short of self-sufficiency. Picture windows appear a staring white, showing that they are radiating heat to the general neighbourhood, and a badly-fitting door shows as a streak of white escaping warm air. An insulated roof shows up bluer than an unprotected one, but even with the insulation there are often features like dormer windows with single-thickness glass and thin side walls that make a startling 'hot-spot' on the Thermovision picture of the roof. The back cover Thermovision picture shows losses from the conventional modern house —every light-coloured patch is an area of heat-loss.

In an average modern house, made to accepted building standards but not specially insulated in any way, it is estimated that three-quarters of the heat put out by fires or central heating disappears almost immediately through the walls, roof, floor, doors and windows. Of this one third is through the walls, a quarter each through the roof and through doors and windows, including ventilation and draught losses, and a sixth through the floor. If your average winter electricity consumption is around 80 units per day, and you pay about 1.3 pence per unit, you can think of these figures in terms of a weekly flight of £1.82 through the walls, £1.40 through the roof, £1.40 out of the doors and windows, and 91p into the ground—not quite lighting the fire with a five-pound note, but very nearly.

If you work out the figures for your own house, you will have a reasonable basis for deciding how much you ought to spend on insulation. Certainly adequate insulation is the first priority for any scheme of house improvement—while self-sufficiency is an admirable ideal, we shall not approach anywhere near it until our houses behave less like radiators to the earth and the sky.

Doors and windows

Some methods of insulation are so simple and inexpensive that you can instal them in an afternoon, and save their cost in a few weeks. Thick curtains over the windows will prevent over half the losses at night, for example. Use a heavy material, line it, and make sure that the curtains are wide enough to give ample folds even when drawn, but don't cover the radiators. Over doors heavy curtains will again give quite good draughtproofing: if you can, set out the curtain rail a little from the actual doorway so that there is room for a person to stand between the curtain and the door. This way he is likely to shut the door before he opens the curtain, instead of opening the door and curtain together and letting warm air out.

For the price of three or four weeks' average heat losses through doors you can put weather-stripping round the sides and top. Phosphor-bronze strip (the springy copper-coloured metal) lasts for years without losing its ability to exclude draughts, but if the door or frame woodwork is warped or irregular, you may find that the strip made of polyurethane foam fills the gaps better. The threshold of the door can be made draughtproof with a synthetic rubber excluder such as Sealmaster, which can be fitted at a number of different angles to seal the gap, according to the type of door-frame. If you are carrying out other alterations to your house, take a careful look at the direction in which doors and windows open: it is common practice in Scandinavia to make sure that the openings face away from the prevailing wind, so that the effects of any small gaps or warps in the frames are minimised. Any other gaps—letter boxes, keyholes, and so on—should be covered with a heavy piece of felt behind the door so that air cannot get in even if the letter flap has been propped open by newspapers or large packets. All these devices may sound trivial, but heat losses are multiplied by draughts.

When you are stopping draughts, take care not to eliminate all ventilation. Apart from the needs of the people in the house to have the air changed regularly, fires and other combustible sources of heat also need a supply of air. If a fire will not burn up properly unless a door or window is slightly open, you need to fit a proper ventilator to provide enough air, otherwise dangerous fumes may blow back into the room. For open fires and stoves there are ventilators that fit on either side of the hearth and take in air from under the floor, thus providing plenty of air without draughts. These are difficult to fit if you have a concrete or other solid floor, and in this case you should fit a ventilator over a door leading into the room with the heater.

Shutters are a very effective way of preventing heat losses from windows at night, and our forefathers used them extensively, making heavy wooden covers that fitted securely over the windows, protecting them from damage and cutting the heat losses by about half at least. Shutters can look very attractive as features on a house wall, but for insulation purposes they should be solid wood, not the louvred type that are often fitted for decoration. These shutters with openings are really borrowed from the hotter countries like Spain where it was essential to have a cover over the windows for protection against marauders, but ventilation was necessary during the hot summer nights. This ventilation nullifies the whole purpose of the shutter for a winter night in Britain. Apart from any other advantages, shutters, if they are well constructed, are very good protection against the casual burglar, if you fit a simple locking device to the catch.

Apart from shutters that fold inwards from beside the windows, you can use detachable covers or screens on any window that is not going to be used much during the winter. Full-length french windows, for example, are often a great source of heat-loss during the cold months. They can be fitted with a cover over the bottom half of the door, which will prevent draughts and a good deal of loss, while not blocking any of the light from the upper part of the door. The cover can be held in place by turnbuckles, if you want to be able to remove it quickly, or made of painted chipboard sliding into two wooden mouldings, one each side of the glazing on the lower part of the door.

For rendering windows permanently less extravagant of heat, double glazing is the only satisfactory solution. Ordinary windows with a single sheet of glass between you and the outside world account for about 10-15 per cent of all heat losses in the house (more if you have very large windows), which may well mean that you are paying between 80p and £1.00 a week in winter fuel bills for the pleasure of having light and a view. Another sheet of glass or transparent plastic arranged so that there is a layer of still air between the two panes will not appreciably interfere with the light or the view, but can easily halve the losses. This saving, between 7½ and 10 per cent of all heat lost, can in 8-9 years pay for most types of double glazing, although you should beware of the claims of some double-glazing contractors who try to suggest that their products will halve your actual fuel bills. The saving is more like one-twelfth.

In Sweden it is quite common to have triple glazing, with three layers of glass, and this cuts down energy loss even more. Such a system could have advantages in this country, particularly for large windows or on an exposed wall. If you live in what the heating engineers call 'severe' conditions—on top of a hill, on the coast or a steep river bank—triple glazing on at least some windows may be a necessary precaution, and many of the systems on the market can be used to add a further layer of glass to conventional double-glazing units.

There are basically two types of double glazing methods which are suitable for permanent installation—sealed units and supplementary frames. Sealed units consist of two sheets of glass fixed into one frame, almost invariably metal, and the units are made in the factory and supplied ready to fix into the frames. Because the sealing is a factory job, the windows are made in a number of standard sizes, and though it is possible to have units made specially, these are very much more expensive than the nearest stock size.

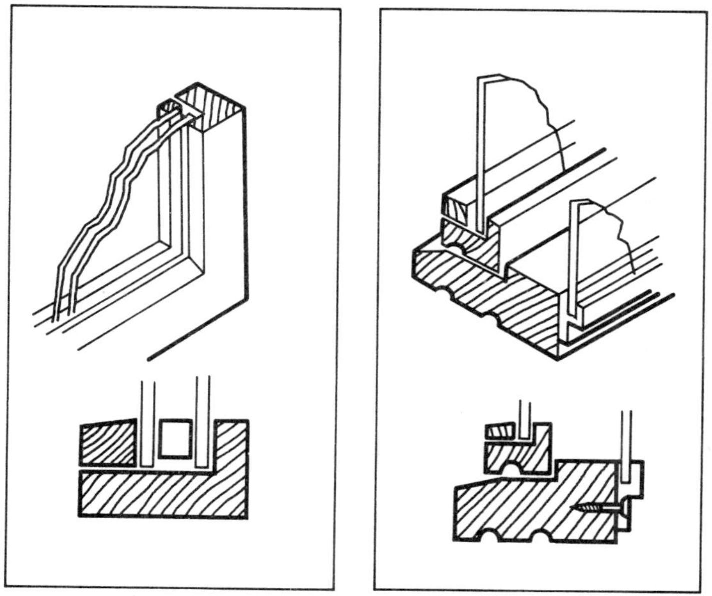

Sealed-unit double-glazing (left) and supplementary frames (right)

Sealed units

There are several advantages with sealed units:

1. Some modern window frames can be used for the units without alteration, although most window frames are too shallow for the rather thick double panes, and may have to be replaced when the sealed units are installed. Units are available to fit standard single-glazed frames.
2. The ideal distance apart for the panes of glass, for maximum heat retention, is about 19mm (although some manufacturers use 12mm spacing), and this distance is easily obtained with two panes sealed into one frame. (The reason for this particular spacing is that closer proximity allows heat loss by radiation straight through both sheets of glass, while wider spacing gives room for the enclosed air to circulate by convection, and heat is transferred from inside to outside by this movement.)
3. Condensation between the panes is eliminated, because specially dried air is sealed into the space.
4. Cleaning sealed units is usually simpler than with other systems, as the sealed inside surfaces cannot get dirty, so there are only two faces to clean, as with an ordinary window. If the modern centre-pivoted windows are used, these can be swung right round to be cleaned on both sides from inside the room.

The main disadvantages of sealed units are:

1. They are no better at sealing the gap between the window pane and the actual frame than conventional windows, and if there is any warping or bending of the window (perhaps by slight movement of the building) a draughty gap can appear.
2. While sealed units can be constructed with the ideal spacing for heat insulation, they are only moderately good for noise insulation, which requires a spacing of 100mm or more to be really effective. Obviously the two sets of requirements cannot be reconciled with any double glazing system, but if noise is a particular problem for you, do not use sealed-unit windows, but have two separate sets of glazing at least 100mm apart. Of course, triple glazing can deal with both problems: a good system is to have sealed units outside, and a further layer of glazing about 150mm further in.

Supplementary frames

Supplementary frames are exactly what the name suggests—another set of window frames constructed inside the main set. Usually the panes are made to slide in the inner set, so that the windows can be opened or cleaned without trouble. Plastic or aluminium frames are usual.

The advantages of supplementary frames are as follows:

1. There is no need to change the existing windows as long as these are in reasonably good condition. This may be important if you have very large windows, or a distinctive pattern (for example, in a period house, where the frames suitable for sealed units would look out of place).
2. Draughts are less likely than with sealed units because there are two lines of defence against air flow. (Even so, it is important for the efficiency of the system that any gaps or draughty places in the existing windows should be repaired before double glazing is done.)
3. Nearly all the supplementary frame systems are cheaper than sealed units of the same size.
4. Well-made supplementary frames (the aluminium type, for instance, with robust fastenings for the sliding panes) are quite a useful deterrent to burglars, as two window fastenings of different kinds have to be forced in order to open a window.

The disadvantages are:

1. Condensation between the layers of glass. Small containers of silica gel can be used to dry the air, but these need reheating from time to time to dry the silica gel again. Small holes can be bored in the frames from outside to allow a very slight air current to pass between the panes and any vapour to disperse: the holes should be plugged with glass wool or some similar porous material to stop insects getting into the system.
2. This type of double glazing means more trouble cleaning, as all the surfaces of the glass can get dirty, and it is usually necessary to lift the internal window panes out of their frames to clean. Insects often get into the space during the summer, and this means extra cleaning.
3. It is difficult to get the two panes of glass as close as 19mm for maximum heat retention. On the other hand, with the supplementary frame system it is usually easy to arrange a 100mm spacing for good sound insulation.

Many firms will do the whole job of double glazing for you, but it is within the capabilities of most handymen. The important point to remember if you are doing it yourself is that the new windows must fit well Gaps or loose-fitting window panes can cause air currents that will nullify almost all the advantages of the double glazing.

You should be able to double-glaze the average house for around £200, using a reliable and robust system, and as this should save you at least £22 a year in fuel bills, it will pay for itself in nine years or less. If you are in rented property, or do not want to lay out so much capital, you can still save some of your heat by fitting one of the cheaper systems based on sheets of transparent plastic stuck to the inside of the existing window frames with double-sided adhesive tape. This arrangement does not usually last for more than about two winters before the plastic begins to crack or tear, but it is very much better than no double glazing at all.

As a parallel to double glazing on the windows, you should consider double doors for your various entry points to the house. A front or rear porch keeps out a great deal of bad weather, and keeps in a lot of heat. The best arrangement, if you can manage it, is to have a porch with its outside door at right angles to the door it protects. This ensures that draughts cannot blow straight through (and it usually gives you more room to open the outside door). In long narrow entry halls you can fit a second door inside the front door, but far enough away from it to make sure that people shut the front door before they open the other one.

Front porches, as they affect the appearance of the house from the road, must be sanctioned by the planning authorities, but you should have very little difficulty in getting permission if you build your porch of reasonably strong material and cover it with tiles or slates to match the roof of your house. Once you have built an enclosed porch you will wonder how you ever managed without it. Apart from keeping out the winter winds, it is a perfect half-way station between the garden and the house, a dump for umbrellas, wellington boots, and all the other bad-weather gear, besides bicycles, prams, and even delicate plants that need to be taken in on frosty nights.

The Macclesfield house goes even further by having a long conservatory covering the south-west wall of the building, so that a large proportion of the area is insulated against extremes of temperature (see page 67).

Ceilings and roofs

About a quarter of the heat losses from the average house go straight up to the sky through the ceilings and roof. Considering that everyone knows that heat travels upwards, most houses are extraordinarily badly equipped to prevent this loss. Perhaps the most striking piece of evidence on our low standards is the figure for 'satisfactory' insulation quoted for roofing in the very latest Building Regulations that came into force in 1975. A roof that satisfies the regulation *thermal transmittance coefficient* of 0.6 still lets out heat at quite a rate. On a frosty day, taking an average-sized roof of about 40m^2 plan area, the structure can easily be losing heat to the outside air at nearly 500 watts, so that you could easily be spending about a pound a week in cold weather just on losses through the roof: as most houses do not even come up to the new Building Regulations standard, the average loss is greater than this.

If you have an old house, it is quite likely that there is no slaters' felt between the tiles or slates and the rafters, in which case the air currents that get in between the gaps in the tiles will be carrying off heat at a tremendous rate, apart from the chance of rain or snow penetrating the roof and spoiling the ceilings below. **Whatever else you do, try to eliminate this leakage.** Taking off the tiles and putting slaters' felt on the outside of the rafters is a long and difficult job, but you can get almost as good an effect by tacking heavy polythene sheeting or builders' paper over the inside of the rafters. Overlap two layers at the top, under the roof peak, and allow plenty of length at the bottom to push the covering well down into the eaves—you may not be able to get at this area with your hands, especially if your roof is rather shallow, but you can usually push the covering down right into the eaves with a soft broom, taking care not to tear the plastic or paper.

This is only a minimum treatment, however. This will eliminate draughts and dirt but will not cope with blown rainwater. For real heat insulation you need a substantial thickness of some porous material that will hold a layer of still air between the interior of the house and the roof.

The first thing to decide, before you do this, is whether you are ever going to use the loft area as a room. Most lofts, if they have not been used for a long time, look very depressing, dirty and completely unsuitable for living or working in, but in many houses they represent a very large useful space that could easily be adapted as a playroom, workroom, or spare bedroom. If you decide that the loft space is usable, you will need to insulate the rafters; if not, it is sufficient to insulate over the joists of the ceiling below, and this is a smaller area, it will be a little cheaper to do the work.

To insulate the rafters, there are several useful materials that can be used either singly or in combination. One of the simplest to install is *semi-rigid slab* made of glass fibre or mineral wool, a porous composition that comes in thick sheets that can be cut with a sharp kitchen knife. The slab has a slight 'spring' to it, so that if you cut it very slightly larger than the spacing between the rafters, you can wedge it into place and it will hold itself up without further support. You cannot put semi-rigid slabs between the rafters if you have covered them on the inside with polythene or builders' paper, so this method is only suitable if you have slaters' felt already fixed between your tiles or slates and the rafters. Be careful about putting the slabs between the rafters if there is no felt over them, otherwise water might penetrate the tiles and be held against the wooden rafters by the slab, thus encouraging rotting.

A better arrangement still is to use glass-fibre insulation quilting between the rafters, and cover up the whole lot with a layer of insulation board or plaster-board nailed to the rafters. This will not only reduce the thermal transmittance of your roof to a reasonable figure, but will also provide a smooth 'wall' round the loft that can be decorated and used to make the space into a habitable room. The glass fibre should preferably be 100mm material. If you do the work yourself—and it is not difficult or heavy work— you should be able to insulate the average house at rafter level for about £50-60, with the possibility that this will save you about a pound a week on fuel bills during the winter, so your work should pay for itself in about two years, apart from the fact that a well-lined loft is half-way to being a useful playroom or workroom.

If your roof space is not really usable in this way (as in many modern houses, where the roof is at such a shallow pitch that the loft is not high enough to stand up in comfortably) you can insulate the house at the level of the joists of the ceiling below. There are several suitable materials for this. One of the simplest methods is to buy bags of loose mineral wool or vermiculite, and spread these between the joists, preferably at about 100mm thick, you can get a level surface by cutting a board to the right depth, with two 'ears' that rest on the joists, and running it lightly down the layer of granules. Vermiculite and mineral wood are non-inflammable, and thus better than an alternative sometimes offered, expanded polystyrene granules. These can be a fire risk if they get into electrical junction boxes in the loft space.

Blanket materials such as mineral wool or glass fibre quilt can be bought in rolls and simply unrolled between the joists. This method of insulating is usually more convenient than using granules, because it is easier to lift up the insulation again if you want to carry out any repairs to the ceiling. The standard thicknesses of quilt are 75mm and 100mm: the thicker the better, from the point of view of insulation. You should be able to cover the average sized loft for about £30, using 75mm glass fibre, so you could easily save your outlay in less than two years.

The wooden joists can also let heat through, and some people lay a further layer of quilted insulation at right angles to the first, so that the joists will be covered as well. If you do this, remember that you will no longer be able to see exactly where the joists are, and it is as well to nail down a few planks over the insulation to show you where you can walk if you have to carry out repairs in the loft. Otherwise you could easily put your foot through the ceiling.

When insulating a loft in this way, you will be cutting off the heat from the house to the whole loft area, and if this includes your cold-water storage tank, you must take special precautions against it freezing. Do not put any insulating material under the tank—this allows a small amount of heat to escape from the room below and keep the water just above freezing point. Preferably continue the insulation right over the tank, so that it becomes effectively part of the warm area below. You can build a simple box in hardboard or plywood to hold glass fibre or vermiculite round the tank, or, more simply, cut slabs of 100mm expanded polystyrene to make a box. Stick the corners together with the cement sold for polystyrene ceiling tiles, but leave the lid loose, so that it can easily be removed if you need to get at the tank and ball-valve for repairs.

You can obviously increase the thickness of any of these materials if you want a higher standard of insulation, and there is no doubt that in years to come we shall regard our present 'insulation' schemes as the bare minimum standard for housing. *Two general precautions* need to be mentioned: first, when insulating your loft, make sure that there is some *controlled* ventilation for the woodwork, otherwise damp may be encouraged. When using blanket insulation over the joists, for example, ensure that the edges of the blanket do not block up the eaves completely, so that air can circulate to the rafters. Second, glass fibre in particular, and some of the other materials to a lesser extent, can cause irritation to skin and lungs. It is as well to use gloves when you are handling insulation, and preferably a dust-mask as well, such as the simple Martindale gauze mask that can be bought for a few pence. This can save a lot of irritation to the throat and lungs from dusty or fibrous products.

If you are in rented property, and do not wish to install a permanent insulation system, you can still save yourself some money in fuel bills by putting a layer of fibrous material over the joists. Use glass fibre quilt in polythene or building paper envelope and take it with you when you move.

At the other end of the scale, to get an idea of the sort of insulation standards that will apply to the houses for the future, the roof designed for the Grants' house at Macclesfield has the following structure:

Solar side

4mm glass
25mm average air gap
0.7mm aluminium sheet
100mm semi-rigid slab
25mm air gap
50mm semi-rigid slab
500g polythene vapour barrier
15mm timber boarding
(fireproofed)

North-east side

Welsh slates
Sarking felt
150mm semi-rigid slab
500g polythene vapour barrier
15mm timber boarding
(fireproofed)

This roof has a thermal transmittance so low that it conserves over twice as much heat as a conventional roof made to the highest Building Regulations standard, and even in the coldest weather the heat lost is not likely to be more than about 200 watts.

Dow Chemicals Roofmate expanded polystyrene insulation boards used on a new estate in Camden

Floors

Suspended timber ground floors (the kind with floorboards over joists) can cause a lot of trouble and heat loss because of draughts getting through gaps in the boards. Air circulation in the underfloor space is essential with this type of floor, to prevent dry rot and other timber diseases developing in the joists and the undersides of the floorboards, but this air circulation can be a nuisance if some of it comes up into the rooms above.

Whatever else you do, avoid reducing the underfloor ventilation. The airbricks and ventilators are very important to the well-being of your house, and it is as well to go round regularly and make sure that they are not blocked and that air can circulate freely under the floor. Concentrate on keeping the floors themselves airtight. Small gaps can be filled with a

The Granada team installing Styrofoam insulation in the floor of the Macclesfield house

mastic sealer, or even with papier mâché made by boiling old newspapers to a pulp and mixing the mass with thin glue. Wedge this into the cracks between the floorboards and where the floor meets the skirting board. Really fine cracks can be sealed with Polyfilla or a similar material. If your floorboards are old and warped, and there are large gaps between them, you can fill these by cutting lengths of lath to a knife-blade shape, smearing them with glue, and hammering them into the gaps, but it is probably better to cover the whole floor with a layer of hardboard.

Over the floorboards, use as many layers as possible to stop heat going downwards into the floor, and draughts coming up. Spread old linoleum or even newspapers over the floor as a first layer, then underfelt, then a carpet (or use a carpet with a bonded underlay). If your budget will only run to a choice of cheap carpet with underfelt, or a more expensive carpet without, choose the underfelt. It will save you money, and when eventually you can afford a better carpet you can still keep the underfelt as additional insulation.

If you want a tiled floor, you can get better heat insulation by covering the floor with 13mm or 18mm wood or flax fibre-board, and a layer of oil-tempered hardboard over this, to support the tiles. Do not use ordinary hardboard, as it will not stand up to the wetting inevitable when you wash the floor. However well you set your tiles in, there are always gaps through which water can penetrate.

Solid floors can be treated in the same way. If you find that you have damp problems due to the age of a concrete floor, give it a coat of one of the proprietary bitumen emulsions used for damp-proofing new floors, before you put any other layers on the surface, or alternatively cover it with builders' heavy-gauge polythene sheeting. If you are renovating such a floor extensively, you can consider putting in a thick layer of extruded expanded polystyrene slab over the concrete, and covering this with a screed of cement. The polystyrene must be the closed-cell waterproof type. This plan is often used for cold stores and other buildings that have to be insulated extra carefully. However, if you do this to an existing floor you must remember that the floor level will be raised, and this will mean rehanging doors etc. Probably the best opportunity to use this technique is when a suspended wooden floor is being replaced by a solid concrete one, so that the levels can be adjusted easily. **The Macclesfield house has 100mm concrete over 50mm of a load-bearing polystyrene called Styrofoam which gives a U-value of 0.36.**

Apart from carpeting, some solid finishes have very good insulating properties. Vinyl cushion tiles have an interlayer of foam that makes them very much warmer to stand on than the ordinary vinyl floorings—these are available from several manufacturers now. Cork flooring is also a good insulator, besides being excellent for deadening sound. Vinyl cushion tiles and 13mm cork are available from about £2.00 per square metre.

Walls

Despite the solid appearance of the conventional brick wall, it is the route through which most of the heat escapes from a house. Even where the brickwork is well done, tiny gaps, multiplied a thousand times, allow a constant movement of air. In addition, when a wall is soaked in rain and then exposed to the wind, the evaporation of water chills the masonry far below the ordinary outside temperature—if you have an exposed east-facing wall it may be behaving as a kind of cold 'radiator' for half the year. It is not surprising then, that up to a third of the fuel bills for the average house are spent on heat that goes straight out through the walls.

The latest Building Regulations standard for heat transmittance of walls is 1.00, which means effectively that the average house end wall, say 4.5 metres square, can still radiate nearly 300 watts on a cold day even if it complies with the new regulations. As very few houses have reached this standard of insulation, the average losses are far greater.

The obvious thing to do is to add on more insulating layers, and these can be put inside, outside, or, in the case of cavity walls, in the middle of the wall.

One of the simplest internal treatments is to hang heavy curtains along the walls, or at least the exterior ones. Our ancestors centuries ago covered their cold and damp stone walls with tapestries and linen curtains from floor to ceiling (and these hangings are now coming back into fashion). Expanded polystyrene and fibreboard make even more effective linings for walls. Polystyrene is cheaper and lighter than most other insulating materials, but can catch fire and give off dangerous fumes and is easily damaged by impact. It is therefore *only* suitable for use if covered, preferably with a layer of rigid material such as plasterboard. When applying the polystyrene to the wall, use one of the recommended adhesives (as sold for polystyrene ceiling tiles), and apply the adhesive to the wall, not the polystyrene sheets. Thirteen millimetre sheets will make quite an effective heat barrier. When covering the sheets, it is best to have a layer of heavy-duty polythene between the polystyrene and the plasterboard.

Thirteen millimetre fibreboard is also a useful insulating material and can be covered by hardboard if a really smooth surface is needed for decoration. Again a polythene vapour barrier between the layers is advisable.

If the wall is damp and therefore cold, try to trace the cause. Damp often rises in a wall because earth has been allowed to pile up at the foot of the wall until it is over the level of the damp-proof course, so that moisture creeps up from the ground. Wherever this has occurred, dig back the earth, clean down the footings of the wall, and fill the trench with coarse gravel or crushed stone, say 50mm in diameter, so that there is no easy pathway for the moisture. Silicone solutions can be injected into the wall to make it more water-resistant, and many firms will carry out this work and give you a guarantee of satisfactory performance for a set number of years.

Where the dampness is simply due to rain and exposure (as, for example, if your house is detached, near the top of a hill, and has one wall exposed to the east wind), it will help matters if you insulate the walls with absorbent material slightly separated from the wall surface. The normal way to do this is to fasten wooden battens, treated with preservative, to the wall, using masonry plugs or pins according to the state of the wall surface, and fix expanded polystyrene sheets to these supports—18mm material is suitable. Then nail another set of battens at right angles to the first set, cover these with heavy-duty polythene sheet to act as a vapour barrier, and use the structure to support plasterboard, chipboard, or hardboard, which will now form the new interior surface.

This may seem a great deal of trouble, and some people feel that such a complicated lining will make the rooms appreciably smaller. In fact, the system only takes about 10mm off each dimension of the room, which is hardly noticeable once the work is done. The main advantage is, however, that it can easily save you as much as 20 per cent of your fuel bills in winter, especially if your existing house wall is a solid one, and this will soon pay for the materials. One minor advantage of fitting this 'false wall' in many old houses is that it gives an ideal opportunity to fit a modern ring-main electricity supply in the rooms with scarcely any more disruption than the insulation itself brings about. The sockets are mounted flush to the false wall, which gives a neat appearance, and the cables can be run round in the gap between the original wall and the plasterboard. Pipes for central heating radiators and similar services can also be run in the gap, although there may be some loss of heat unless the pipes are lagged carefully—as there is no point in heating up the insulated area.

One reason for putting the polythene vapour barrier in any system of internal insulation is to prevent, as far as possible, water vapour from the warm interior of the rooms getting into the colder exterior parts of the wall. In cold weather there is a danger that condensation could take place in this area, so you will need to seal carefully around the wiring for instance. An extruded silicone rubber sealant would be ideal.

If you have cavity walls, one effective way to insulate is to fill the cavity with foam material which will stop air from circulating and thus carrying away heat. The normal foam used is urea-formaldehyde plastic foamed up as it is injected into the wall, or loose mineral wool blown into the cavity through holes drilled in the outer skin of the wall, and packed to a uniform density. This is a popular means of insulation, because all the work can be done from the outside and there is very little disruption of the house, and, if properly carried out, cavity insulation can reduce the heat lost by a wall to almost a quarter of its previous value.

However, there are complications. Regulation C9 in the Building Regulations 1972 says effectively that nothing must bridge the gap in a cavity wall except the specified building ties that help to hold the walls together. This is not an arbitrary prohibition, but is designed to make sure that moisture cannot creep from the outside leaf of the cavity to the inside wall and thus make the interior of the house damp. While properly applied mineral wool or plastic foam will not permit water to pass, some firms who set up to carry out cavity insulation when it was first introduced had neither the experience nor the thoroughness to make a good job of filling the cavity. The gaps that they left acted as channels for water to pass between the walls, and this sometimes made the walls less weatherproof than they had been before treatment. Many local authorities, worried by the apparent poor results of cavity insulation, decided that they must enforce Regulation C9 and refused permission for further insulation work. So despite appeals from the Department of Energy for people to 'Save It', many people were prevented from having cavity insulation carried out.

The situation is now somewhat clearer, and in most areas you can get permission (or a waiving of the requirements of Regulation C9) if you agree to use the services of one of the firms experienced in the work, as expressed by their possessing the so-called Agrément Certificate. While it is probably true that a competent handyman, given the right materials and the time to carry out a thorough job, could insulate a cavity quite adequately, it is most unlikely that you will get Building Permission to do the work yourself. If you insulate your cavity walls without permission you can be fined quite heavily.

The normal procedure for filling the cavity is to drill a series of holes in the outer leaf of the wall. For urea-formaldehyde foam these are usually 25mm holes at about 0.9 metre spacing in both directions, for mineral wool the holes are 63mm in diameter and spaced at about 1.8-2.7 metres. In either case the insulating material is injected in from the

One of the methods of insulation tested at Macclesfield. An inner layer of glass fibre and an outer one of expanded polystyrene, covered with breather paper

lowest holes until it begins to appear in the next row up, and then so on until the cavity it full. The holes are filled with mortar tinted to match the wall, and the job is complete. It usually takes about one day for the average house.

For new houses there are several treatments that can be introduced while the cavity wall is being built, and if these consist of sheets of waterproof as well as insulating material, the requirements of the Regulations can be met.

But it is best to protect the wall externally—with slate or tile hanging or other cladding—and cope with the problem at source.

For the outside of the wall, there are a number of treatments that can add to the insulating qualities of the building, and most of these are suitable for use with either solid or cavity walls. However, with any covering applied to the external faces of your house, you will have to consider the effects on the visual quality, and you will probably have to get permission from your local authority, because these treatments can sometimes be considered to spoil the visual amenities of the neighbourhood. If you live in Bath, for example, and suggested that you wanted to cover the face of your Georgian house you would probably meet with some disapproval, however much this improved the insulating properties, and certainly if you live in any sort of conservation area you should check carefully with your local planning officer before making any major changes to the look of your house.

One thing that you can do without permission, however, if you have a brick-built house, is to examine the pointing carefully, and renew any shaky mortar. A great deal of the heat loss from brick houses is due to the flow of air through thousands of little cracks and holes, and repointing will not only make your house more proof against damp, but warmer. Pointing is a tedious task, but you soon get used to the technique and it is something that the amateur can do a square metre at a time if leisure is limited. While looking at the pointing, examine the state of the fillets round doors and windows, and plug up any gaps in these that could be letting heat out.

The best method of external insulation, undoubtedly, is to cover the walls with an insulating material such as mineral wool or glass fibre, and then cover this with a weatherproof finish—rather like the interior insulation method, but designed to face the weather. As the Grants' house had been built with solid brick walls that were quite inadequate as heat-barriers, it was decided to build a complete insulated surface over the brick face.

This was done by fixing battens to the walls at 650mm intervals, and placing panels of insulation between these. As the house is experimental, various types of insulation are used—50mm glass fibre

Semi-rigid slab provides extra insulation between the leaves of a conventional cavity wall

Detail showing external insulation of existing brickwork and base treatment at ground level

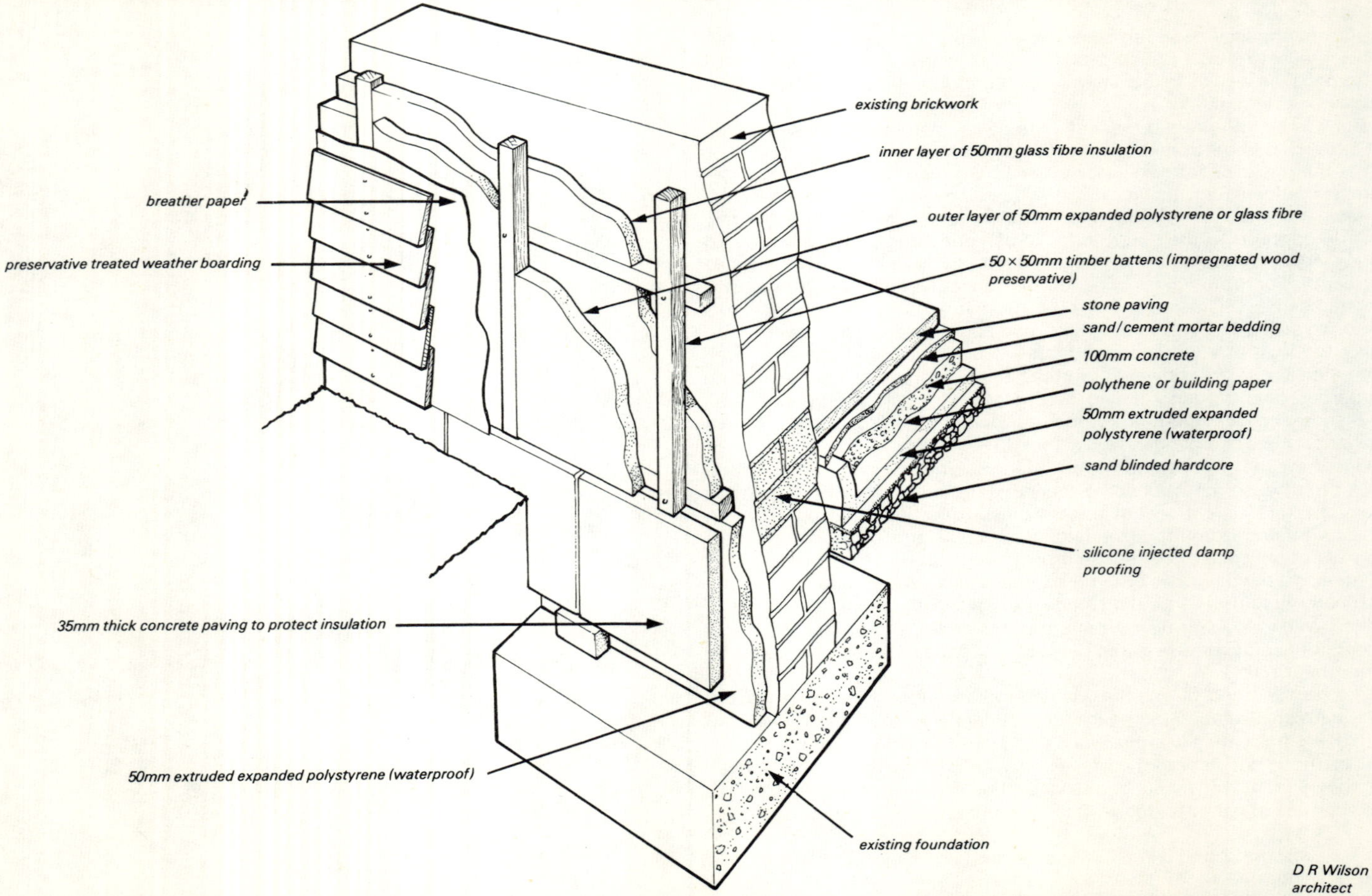

D R Wilson
architect

blanket and 50mm expanded polystyrene slab, 100mm ordinary loose glass fibre on another face and 100mm semi-rigid mineral wool slabs on a further one. While the polystyrene stands up better without much support, the Grants and the Granada team found that it needed more precision in setting out the battens, as it has very little elasticity and cannot be squeezed in if the battens are slightly too close together at any point. The glass fibre, being soft, could of course be deformed to make it fit almost any spacing of the battens, but it needed more care to make it stay up in place until the next layer of the wall was put on. The easiest material to apply proved to be the semi-rigid slab. The performance of all these materials will be checked over the next ten years, as part of the Granada Television experiment in self-sufficiency.

Over the insulating slabs there is a layer of water-resistant breather paper (Sisalcraft No. 40), covered by sawn tapered section weather-boarding that forms the actual outer face against the weather. (Suitable timbers for external cladding are listed in the Building Regulations. Should you think of following the example of the Macclesfield house, check with your local planning authority that timber cladding would be acceptable in your area, and ask them for advice on the type to use. In many districts timber cladding is not encouraged.)

The final result of this protective build-up is a striking demonstration of the quality of insulation that could be achieved if we designed our houses rationally. Instead of the Building Regulations permitted thermal transmittance of 1.00 for walls, the Macclesfield design has a coefficient of about 0.3, over three times as effective at keeping in the heat. Compared with the original walls of the house, which were solid brick with an insulation value far below even the Building Regulations standard, the new walls are about six to seven times as efficient as heat barriers. In terms of savings, this means that the insulation and cladding on the walls alone will eliminate about 20 per cent of the fuel costs that might be expected for a house of this size. Combine this saving with those from the heavily insulated roof and the double windows, and it is clear that Geoff and Lynn Grant are likely to be paying only a fraction of the fuel bills they had to face in their previous, conventional house.

A do-it-yourself energy saver

Build your own removable shutters for your windows as shown below. They are to be hung inside. Lynn Grant is going to cover hers with material and hang them on the wall during the day as decoration

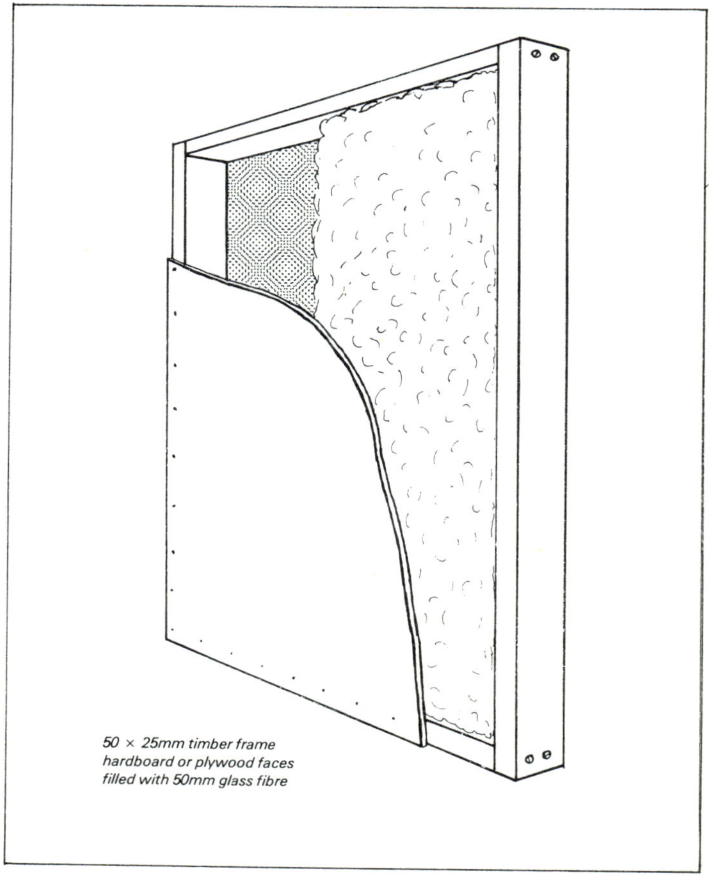

Chapter 12
Making Use of Alternative Technology

by Brian Trueman

As Donald Wilson keeps saying: don't let anyone tell you it's easy. Making a simple drawer isn't *easy* either, nor is wiring in an extra electric point, nor is mending a leaking pipe. But many a person has found he can do all these things when he didn't know he could—and has pride and satisfaction from the independence that doing them has given him. Using sun and wind and water is very little more difficult than these jobs—it needs the will, it demands thought, concentration and care but in return it provides not only a return on capital investment and a saving on the fuel bill but the romantic satisfaction of using the natural forces that 20th century civilisation has very largely lost.

So, if you want it easy, carry on watching the least demanding television and taking the dog for a walk; if you want to have fun and help yourself—read on!

Using the sun

What we should point out is that if you want to save fuel and money the magic word is insulation. You'll get much more money back in a shorter time by investing in a sensible thickness (150mm if you can) of loft insulation or by having your cavity walls filled than you will from a solar panel or two. But then again insulation is fairly tedious stuff—worthy but dull—and we wouldn't like to discourage you from experiencing the romantic thrill of bathing in solar heated water, even if it isn't cost effective!

Flat plate collectors

Take a look at the exploded drawing on the next column and you'll see that solar panels are simple enough. There are four main elements:

i the absorber plate
ii the container
iii the insulation
iv the glazing

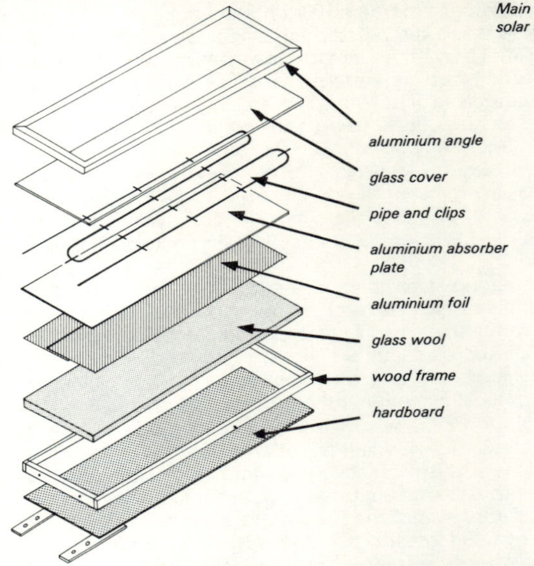

Once you see the principle you can start inventing your own. Mean time you'll see that there are umpteen variations on the main element—the flat plate collector.

1. It can, like the ones on the South wall of my own home, be based on a radiator.

   I bought three, each 3 metres by 0.6 metres through an advertisement in Exchange & Mart and they cost about £7.50 a square metre and as far as efficiency is concerned there are few things better.

2. It can be home-made by clipping, wiring, soldering or gluing copper tubing onto a sheet of aluminium. What diameter tube at what intervals on what thickness of sheet could be the subject of a good few years experimentation but try 15mm tube at 150mm centres as a start. You can either bend it into a serpentine shape or you can use plumbers' capillary right-angle and T-joints to make a sort of 'ladder' pattern with vertical tubes between top and bottom feed pipes.

3. It could be a variation on the 'open' system of the Macclesfield house. The difference between this application and that on the Grants' house would be that this would be a smaller individual unit in its own 'case' rather than an entire roof covering.
4. Whatever form of collector you settle on, it will have to be given a matt black finish. I'm using engine paint for its heat resisting qualities, others use blackboard paint.

The unit will have to be held in some kind of frame or container and it would be ridiculous to lay down any hard and fast rules over the construction of this element. Once its function is defined it's up to the individual constructor to choose his own materials. So

1. It must hold the absorber plate, the glazing and the insulation.
2. It must protect them from the weather and be weather resistant itself.
3. If glass is used as the transparent cover it must be rigid enough not to flex so much as to crack the glass.
4. It must include anchorage devices such as metal straps or brackets to hold the panel securely wherever it is mounted.

You can see that you could make the case out of almost anything—steel, wood, aluminium, glass reinforced plastic (fibreglass plus resin, that is), concrete or solid silver!

Similarly the insulation is merely needed to prevent the collector plate losing heat from the rear surface or from the sides. Some constructors use a layer of aluminium foil to reflect the heat from the upper surface, some don't bother. Given that, though you could use mineral (rock) wool, glass fibre, polystyrene bead board—whatever you find most convenient most available and cheapest. I am using an extruded expanded polystyrene board, 50mm thick, which is somewhat more expensive than most but which is waterproof, tough and reasonably rigid. That means I can do without a base to the tray so that the overall cost is reduced.

Finally the glazing—and it will probably have to be glass. There are a few tough plastic films that will resist ultra-violet light but they are not readily available yet. The glass can be 4mm (32oz) horticultural glass, which is a good deal cheaper than window glass and almost as good and a single layer is all you really need, unless you particularly want very high temperatures in really sunny weather. To hold the glass in place you could rebate the frame and putty it in like a window, hold it down with alloy angle, 'glue' it down with silicone rubber using a few clips. Just make sure that rain-water neither gets in, nor collects where the glass meets the frame.

How much and where?

It's generally agreed that four square metres is about the minimum amount of flat plate area to provide you with hot water. If you have more you'll be better off in winter but you may be embarrassed with too much hot water in summer—and, of course, the bigger the collector area you have the bigger the storage volume you'll need. As a rough guide you'll want 45 litres of storage capacity to every square metre of collector.

Chapter 2 has explained the principles for locating the collectors. The roof is an obvious place but you could also use them as an awning over the windows and get two uses for the price of one, you could mount them in the garden and use them to heat the house, the greenhouse or the swimming pool (if you should be lucky enough to have such problems!). I am pivoting my collectors on my South wall so that they'll be nearer to vertical for the low winter sun and tilted upwards for the summer. Whatever else you do, fasten them *securely* at upper *and* lower edges. You can slide metal straps under the slates and fasten them to the rafters or purlins. There's no one answer so use your ingenuity but look out for shadows and the planners!

Storage

It's one thing catching the sun's energy; it's another thing storing it and the collector can only be part of a system, as of course is your oil-fired/gas-fired boiler.

As far as getting free hot water is concerned though, it comes down to two ways of moving the water around and two ways of storing it and we'll deal with storage first. Figure 1 is a domestic hot water tank

Figure 1

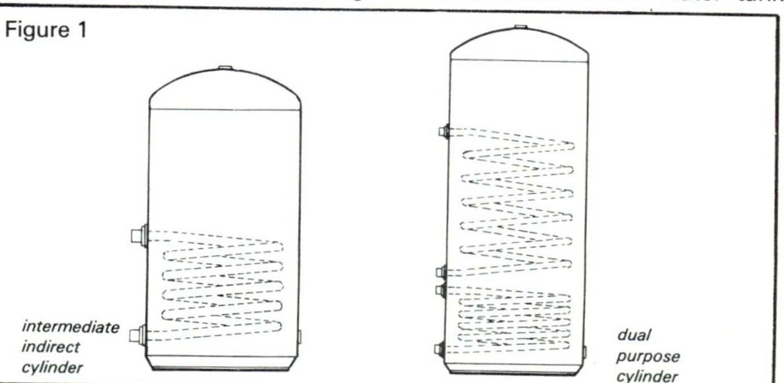

Upper floor layout

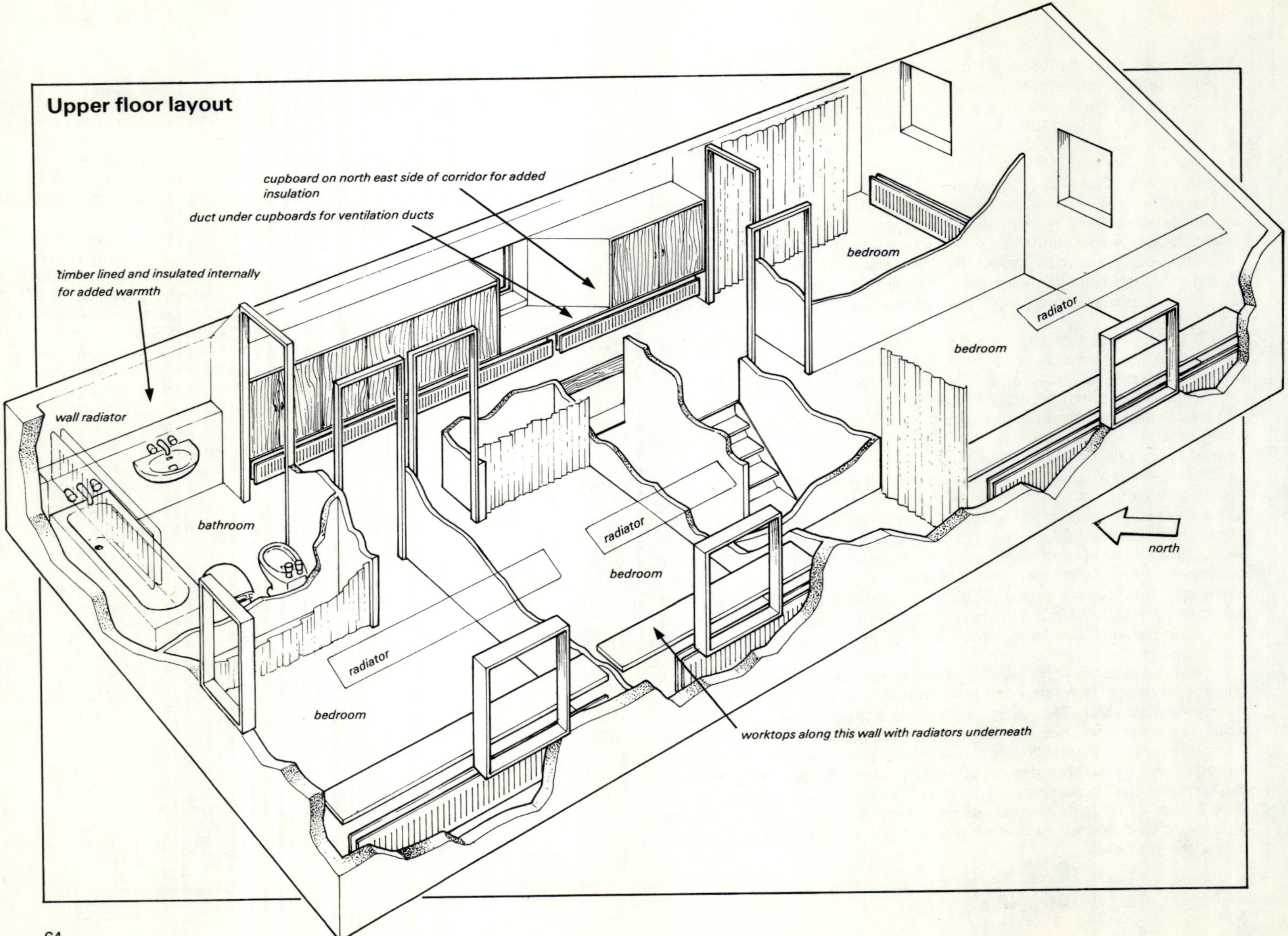

Lower floor layout

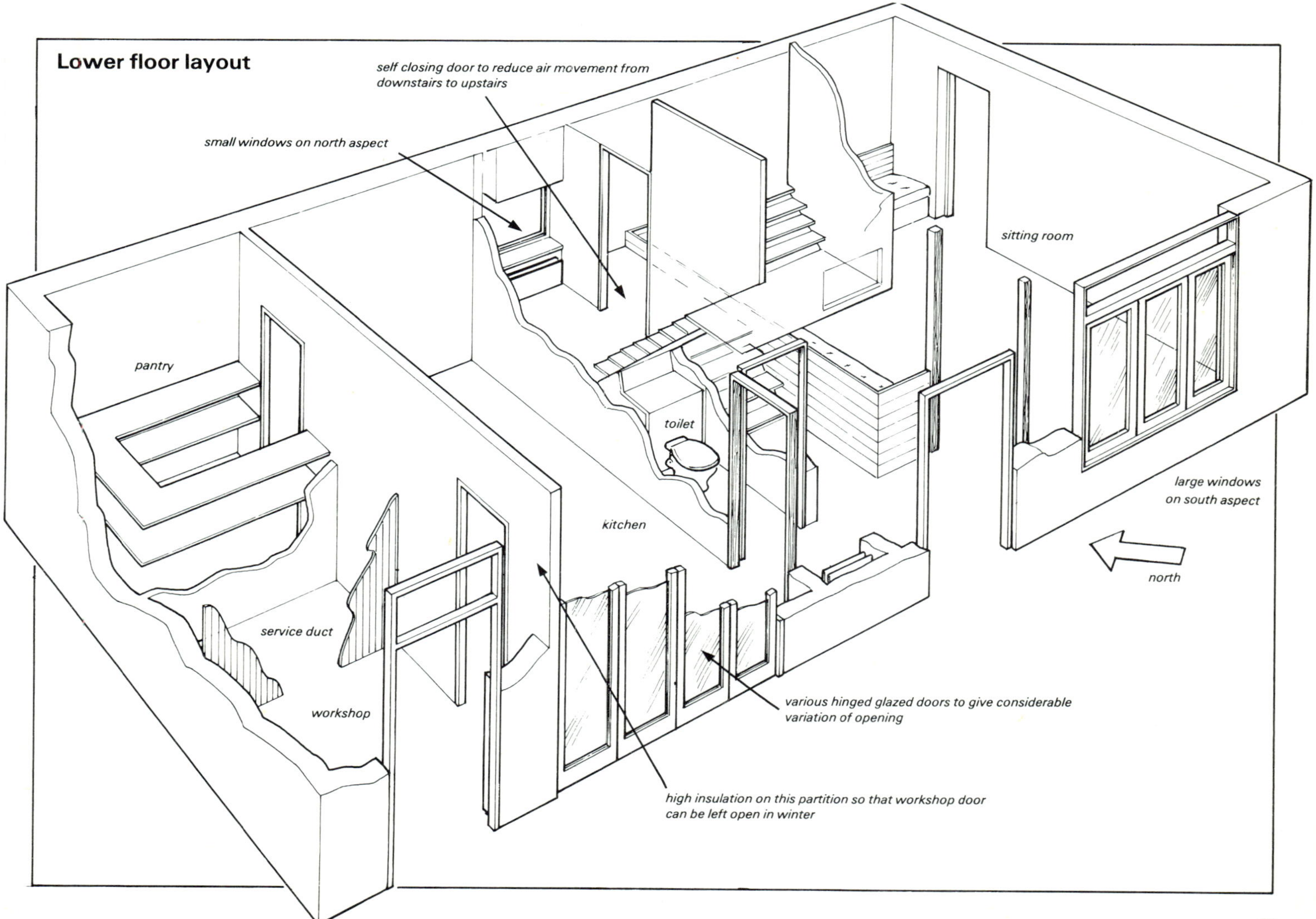

Layout of mechanical ventilation system with heat recovery characteristics

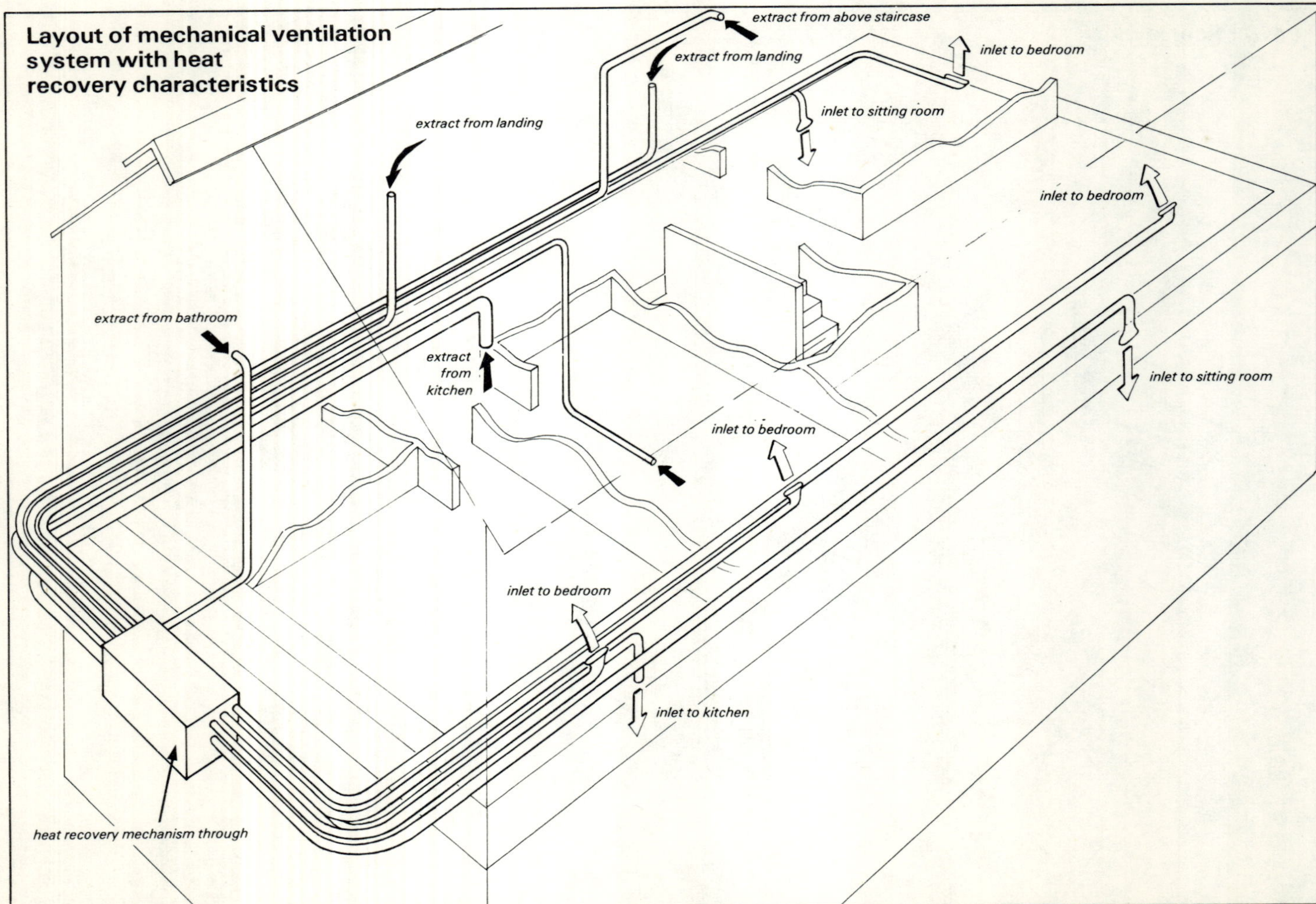

South west elevation

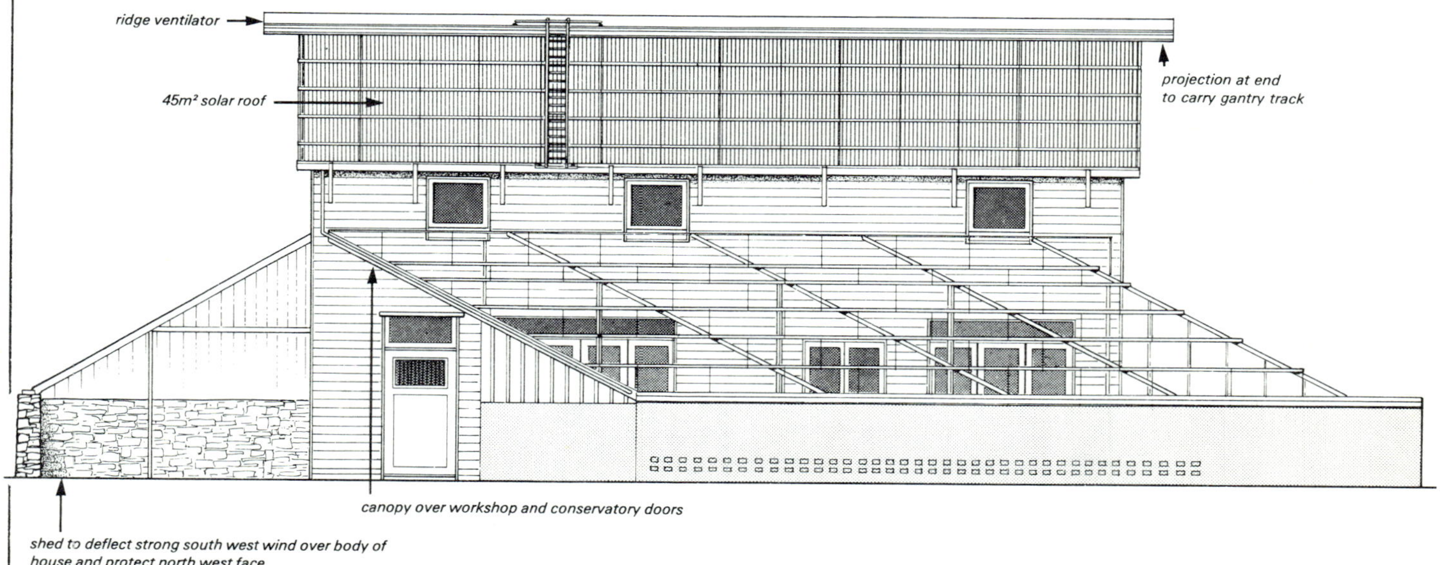

D R Wilson
Architect

available commercially through most plumber's merchants. Instead of having just one heat exchange coil fed by the central heating system or the back boiler it has that one above and a second one, fed from the solar collector, below. In operation the water in the tank whether it's heated by the sun or by fossil fuel will rise so that the lower half of the stored water will be the cooler. That being so it is more likely that the warmed water from the solar panels will have some relatively cool water to which it can pass its heat. As water is drawn off for use, cold water will flow into the bottom of the tank where it may be pre-heated by the solar coil before rising to collect the higher grade heat available from the upper heat exchange coil.

Nevertheless this system does limit the usefulness of the collector since only in summer are the temperatures in the collector likely to be that much higher than those in the tank at the right times.

Because of this a system which uses an intermediate tank is preferable—it's the approach I am going to adopt and it's illustrated in Figure 2.

Figure 2

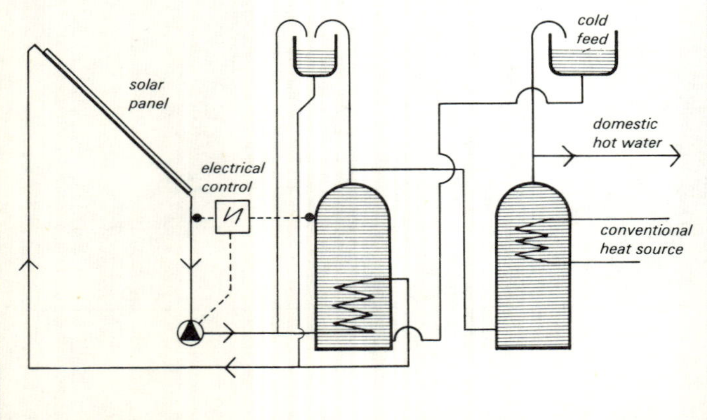

Here the domestic hot water tank takes its cold water supply, not direct from the mains but from a separate tank which contains a heat exchange coil connected to the solar collector. This means that the whole of the capacity is available to receive heat from the panels, using the boiler will not affect the temperature of the solar storage water and the usefulness of the panels will be extended into winter when there may not be much sunshine but where what little there is can lift the temperature of the cold water coming into the domestic hot water tank by a few fuel-saving degrees.

Moreover it is possible to make an indirect tank yourself. A galvanized steel tank may be used with a heat exchange coil fabricated from copper tube formed into a rectangular spiral by using right angle capillary joints as in Figure 3. There's a lot of hassle about the area of heat exchanger you'll need in relation to the amount of water you're storing and in theory it should vary with the sun. Sad to say it can't so I'll make an inspired guess and suggest 1.1 square metres of heat exchanger for every 250 litres stored. That according to my shaky maths, will mean 12 metres of 28mm tubing in a tank of that capacity.

Figure 3

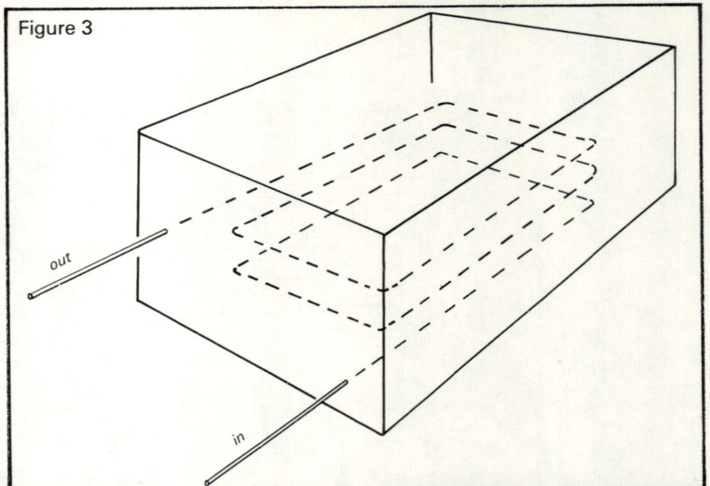

There is, however, still the problem of getting the solar heated water to the heat exchanger and there's more than one way of doing this.

The Thermo-syphon

The simplest and cheapest—though not necessarily the best is the thermo-syphon. If the panels are mounted on a wall, on the ground or as an awning it's quite likely that the flat plate collector will be lower than the storage tank. In that case, as the water in the collector is warmed by the sun it will rise up one leg of the connecting piping and flow round the heat exchanger in the storage tank by convection. As the water gives off

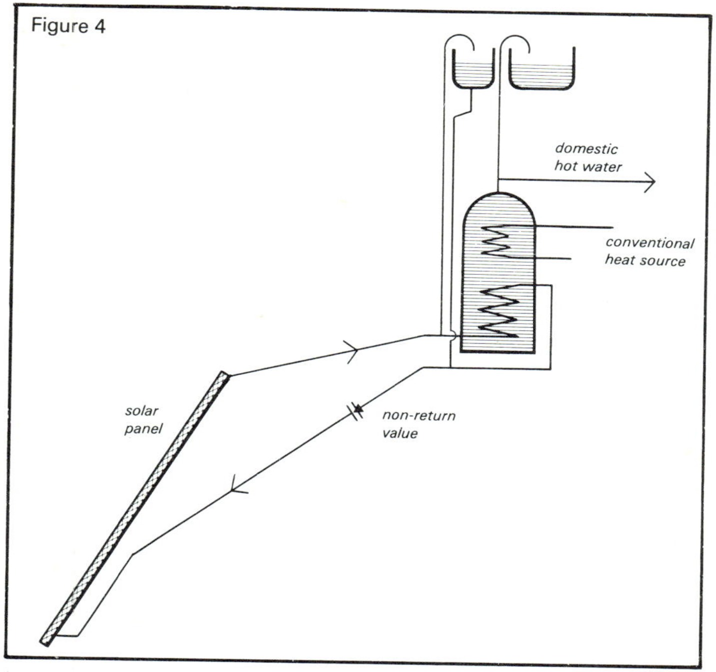

its heat to the cooler, stored water it will remember its O-level physics, became denser and drop down the other leg of the piping.

As the difference between the heat exchanger end and the panel end diminishes the flow will decrease and will finally stop altogether unless the stored water is used and replaced by fresh cold water or the temperature at the panel rises with a higher intensity of sunlight.

There are some important points to watch about this system:

1. The piping will need to be more like 28mm than 15mm to enable the water to flow freely.
2. You'll need a non-reverse valve on the return leg—which isn't as painful as it sounds! Like all the other bits you can get them from plumber's merchants.

The pumped system

A pumped system is more elaborate and needs a source of electricity at 240 Volts both for the pump itself and for the electronic device which controls it. At the same time it will be more efficient, more controllable and will allow you to use 15mm pipe or smaller, which will win back some of the money you spend on the fancy bits, and it will heat your stored water to the same temperature in about half the time. For what it's worth it's what I intended to do and you can see the general arrangement in Figure 5.

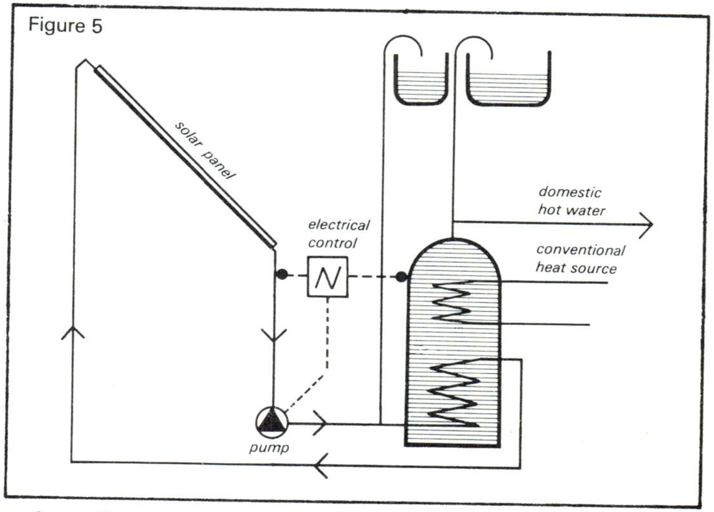

According to research work done in Australia, whatever pump you use should be capable of circulating 40 litres of water an hour for every square metre of collector area. So if you've got 4 square metres you'll need it to propel 165 litres an hour.

Most central heating pumps will shift much more than that but that gives you the chance of putting in a gate valve so that you can adjust the flow for the best performance from your panels. Obviously the more sun, the hotter the panel; the hotter the panel the more flow you need. Don't forget too, that the higher the pump has to lift the water the smaller the flow will be. Your plumbers' merchant will advise you.

You can't leave the pump running all the time, of course. If you did you'd find that you were taking heat *out* of the hot water tank and radiating it to the cold night sky from your solar panel. So a control system is needed. You can buy them for £35 or thereabouts from the solar panel manufacturers, or you can build one yourself for £8-£10 from a circuit design that will cost you 25p and a s.a.e. to BRAD, Eithin-y-Gaer, Churchstoke, Powys.

I'm all right with a soldering iron and I've built some hi-fi so I'm making my own.

What the unit does it to compare the temperature at the panel with that at the storage tank. If the panel is hotter it switches the pump on, if it's colder it switches off. Isn't science wonderful!

A few more points to remember:

1. Make sure your pipework slopes upwards just enough to let any air rise out of the system.
2. As was suggested in Chapter 2, unless you drain the whole system every October—which is silly—you must add 20% ethylene-glycol to the water in the panel to prevent freezing.
3. Be very careful about mixing metals. An aluminium panel with copper piping will soon develop holes because of electrolytic corrosion. Use polypropylene or steel tubing and consult one of the firms that make chemicals for central heating systems about an anti-corrosion additive.
4. So that the heat exchanger is always full, connect the input from the solar panel to the lower connection. That's the opposite of what you'd do in a fossil fuel system.

Geoff Grant on the gantry ladder glazing the roof. This ladder will be used to facilitate maintaining the roof throughout the year

Windmills

If you blunder when you're making or installing a solar heating system it may leak, it may not work at all and I suppose it could even blow up. But make a mistake with a windmill and someone could be killed.

That's not an attempt to put you off, merely to encourage thought and care. We have a book list in my office at Granada TV, Manchester 3, which includes publications on most aspects of alternative technology including windmill design. The more you swot up on the whole field of study the more likely you are to succeed. It can become a fascinating obsession. I should know!

It's useful to realise how much power you can get from the wind. A useful formula is $0.0002 \times V^3 \times D^2$, which, if you can cadge a calculator for half an hour, will give you the shaft power of a windmill of given diameter (D, in metres) in a given wind-speed (V, in metres per second). Our windmill at Macclesfield is 4.5 metres in diameter and in a wind of 5 m/s we'll get:—

$$.0002 \times 5^3 \times 4.5^2$$
$$= .0002 \times 125 \times 20.25$$
$$= 0.50\text{kW}$$

Someone wrote to ask for advice in building a 5kW mill. You'll see that to get 5kW in a 5m/s wind he'd need a rotor diameter of about 14 metres which wouldn't make him too popular with the neighbours!

Nevertheless half a kilowatt for half the year, which is what we think we might get at Macclesfield works out at an average of 6kWh every day.

There are, however, several problems. To get into the faster, smoother airflow our 4.5 metre is over 9 metres up, but it must be lowered for servicing. Then again because the power of the wind increases with the *cube* of its speed it hasn't only to be efficient to make the best use of light breezes, it also has to be able to live in strong gales which means that its blades must feather—turn edge on to the wind—so that it doesn't turn so fast that it explodes!

The generator needs something like 1000 revs per minute to turn out useful current but we daren't let the rotors do much more than 100rpm—that means gearing up at least 10 to 1.

None of these conditions is easy to satisfy. All I can offer you is blood sweat toil and tears—and the magical, child-like delight you'll get when you see your windmill working! So let's go through the problems again.

The mast

We bought a second-hand, 9 metre lamp-post from the Greater Manchester Council for £7.50. Transporting it to the site, 32 kilometres away was another £20 but even so it was dirt cheap compared to a steel lattice tower or a new steel tube of similar diameter. It stands on a concrete cube about 1 metre in all dimensions. A little way up from the bottom of the pole we drilled two 25mm diameter holes opposite one another. A steel bar passes through these and through two hefty brackets which are bolted to the concrete either side of the pole. This way it can hinge up and down.

At equal spacings around a 9 metre diameter circle with its centre at the base of the pole we dug anchorage points. Each one starts with two holes at an angle—like legs astride—1.5 metres deep, 250mm in diameter. An inverted 'V' was made out of 15mm steel rod, each leg was shoved into the bottom of one of the holes then concrete was shovelled in. That left us with a metal 'arch' above ground at the three points. We had stainless steel yacht rigging cable (2700 kilos breaking strain at 25p for 30cm) made up to length with an eye at each end. One was fixed with a 'D' shackle to a lug welded to the pole so that it would be just below the lowest point of the rotor, the other was anchored to a cable tensioner which in turn was clipped, via a carbine hook, to the anchorage point. A second set of cables went from the ground to a point about a third of the distance down from the upper lugs to the ground. All these bits, like many others, came from sailing equipment retailers or ship's chandlers. For the windmill builder, these places can be a source of inspiration!

A final touch of sophistication was to buy a geared winch to haul the pole upright since once the windmill gear was fixed to the top of the pole the all-up weight was about 8cwt and top-heavy. That set us back £130 but the only real substitute we could think of is half a dozen strong friends on constant alert!

The rotor

At Macclesfield we'll have two types of rotor. One is based on the Cretan sail-type. It's the easiest kind to make, cheapest—since there's no welding or machining, and good to look at. Furthermore it won't over-rev because when the speed of rotation reaches a certain relationship with the wind-speed, the sails start to flap, lose their aerodynamic efficiency and slow things down again. This auto-breaking effect, however, will depend on the load it's driving and it would be as well to anchor the apex of each triangular sail to the next 'spoke' with elastic rope.

The Macclesfield Cretan windmill

Fixing the mill to the hub

Excessively high winds will then be spilled as the rope stretches. In working out what strength of elastic to use make sure you begin with the weakest; it's far better to have a less efficient mill than a smashed one! In all other parts of the design, to quote Donald Wilson again, 'If in doubt, make it bigger!'

Our other rotor is a twin bladed affair with aerofoil section blades.

I asked Dr Bill Cartwright from the Department of Mechanical Engineering at the University of Manchester Institute of Science and Technology (UMIST hereafter!) if he would design a 4.5 metre diameter device to give us one horse-power in a 5-6m/s wind. Within a week he came back with a small treatise on windmills, a design and some constructional ideas! There isn't room to go into details here but we have copies at Granada Television.

Feathering these blades was, as they say, something else. As we go to press our friend John Winn is still finishing off the construction of the system! Details will be available in the second group of programmes and eventually in printed sheets from the 'House for the Future' office. But it still isn't tried. Moreover just working out the strength and length of the springs that allow the blades to turn edge on to excessive winds involved calculating the centrifugal force, the wind thrust and the feathering torque, the spring rate and various moments of force. Not only was that beyond me, I didn't know what most of them were before I started! Again Bill Cartwright provided the answers—and we'll pass them on when we've seen how things work out.

The generator

I decided that it would be better to store the output from the windmill as heat rather than as electricity. If you haven't any mains juice at all your decision will no doubt be different but the fact is that far more electricity

The hub

is consumed in heating homes than lighting them and that is where most savings can be made. Moreover, a tank of water and an immersion heater, even a specially-built low-voltage one, costs a good deal less than a collection of batteries. You can take your pick!

The first step was to find a generator that would put out a respectable amount of electricity at low revs. The answer was a lemon—there doesn't seem to be one! In the end we borrowed an AC203 alternator from Lucas/CAV.

It's designed for passenger vehicles and puts out 60 amps at 24 volts (1440 watts) at 1500rpm. Whether it ever gets to 1500 revs is a different matter! Like most generators it consumes twice the horse-power it produces and I reckon that our two 24 volt 600 watt immersion heaters will only be fully fed when the wind reaches 8.5m/s, and there aren't many hours in a year when the wind blows that hard. We may have to settle for a smaller alternator feeding just one of our immersion heaters. Another reason for that decision could be that the generator we borrowed would cost over £300 to buy!

The gearbox

I said before that the alternator has to turn at about 1500 revs per minute to produce its rated output. I also said that our windmill would feather at 100rpm. The answer is a gearbox stepping up the output from the shaft by fifteen times (ours is actually 16:1 for convenience). That was put together for us by a big transmission manufacturer and cost over £100! The alternative, however, was a complicated set of v-belts and pulleys with shafts and bearings. It would have taken more space, much more time and may not have been much cheaper.

You can see how all these elements fit together in the photograph on page 74. The shaft is converted from an old scrap Rover 90 back axle (£10). That gave us the basic metal work and some spare parts. We kept and relined the brakes for obvious reasons. The whole unit, including the rotor and the tail can turn horizontally on a large thrust bearing to face into the wind and John Winn is working on a device that will allow a spring to pull the tail into the same plane as that of the rotor (instead of at right angles to it) when the wind gets too high (say over 15m/s). That way the rotor will be edge on to the wind and will neither revolve nor blow to bits!

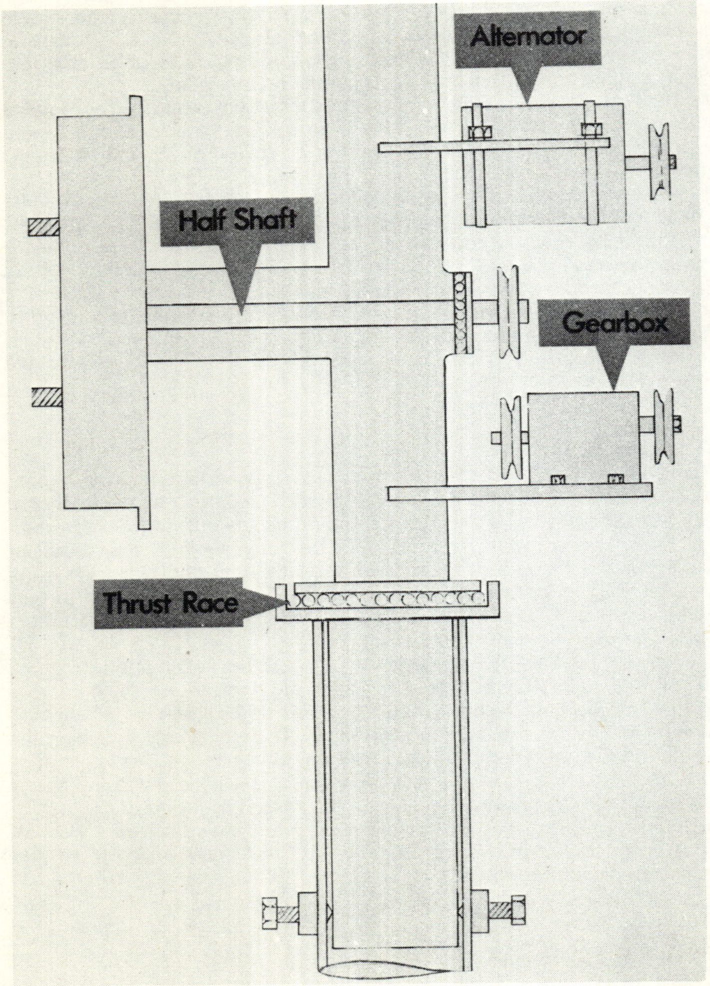

Commercial windmills

One way and another, the windmill is one of the trickiest aspects of alternative technology and it's easy to see why people often settle for a commercially built unit. If you're thinking of investing in one, however, be careful—

1. Make sure the planners will let you have your windmill!
2. Check the wind speed at which the mill gives its *rated* output. There wind to do it and in the Macclesfield area winds like that blow for about 7% of the year. At £2000 plus for the mill, is it going to be worth it? That's why the French 'Aerowatt' windmill producing 4kW in a 5m/s wind costs £15,000!

That's also why the work Dr Peter Musgrove is doing in the Department of Engineering at Reading University is so important. His windmill is a development of the Darrieus windmill. His version looks like an old 'H' TV aerial. See page 32. The blades are straight aerofoil section 'planks' mounted vertically at the end of the horizontal area and the horizontal arm is pivoted in the centre so that it revolves on top of a mast. The blades are connected to the horizontal member with hinges somewhere *below* the half-way mark so that as they spin, their top throws outwards, being restrained beneath a certain speed by cables which run over small pulleys, down into the vertical mast where they are attached to a spring. You can see that when the mill starts to go too fast, the blades will tilt towards the horizontal and lose efficiency. The mill will slow down, the spring will pull the blades towards the vertical and the mill will speed up again.

It offers all the advantages of the Darrieus mill *and* the benefits of a controlled system and very easily and inexpensively made blades which have neither taper nor twist.

Windmills are a challenge—but they're more fascinating and exciting than almost any other collector of free power. And they *can* be built by the amateur with care and patience and thought. Good luck!

Other technology

The Macclesfield house incorporates some pretty fancy devices—most of them are nothing new to industry but they are unusual things to find in the home and, by domestic standards, expensive. Using the sun for space heating as well as hot water, for instance, is a whole new ball-game! In our case it means 5000 litres of water in two tanks, £650 worth of control equipment, £300 worth of piping and a £1000 heat pump.

In this area, though the house is still first and foremost a home, it has become an experiment beyond most people—so I won't go into the details of these installations here. Those who want more details so that they can try their own experiments—use the list at the back and then you can always write to me. If I can't help I'm almost sure to know some person, organisation or manufacturer that can.

It's for other reasons that I haven't talked about methane production. The methane digester is one of the great symbols of alternative technology, as well it might be since it uses bacteria to eat up all sorts of waste—including human—and gives in return methane gas for power and rich compost for growing.

What put us off was that the more we looked into methane digesters (not a very nice occupation, anyway) the fewer we found that actually worked. Lawrence Hills of the Soil Association is one of the great enthusiasts for methane but even he doesn't believe that they are practical for small urban domestic applications. The truth is that a family of four doesn't have enough vegetable, animal or human waste available to produce a reasonable volume of gas. What's more, in winter some of that gas may have to be used to keep the system at a working temperature and that reduces the effective output still further.

Once again there are books which will help you if you are keen to try— for instance, *Methane Power Plants* by L. John Fry.

Self sufficiency in the garden

All I can say about growing your own food here is that the same principle, that of knowing *why* you're doing something, applies here. As it does in the use of sun and wind, in fitting out the kitchen of the Grants' house and in every detail of the whole design.

Our adviser on the garden at Macclesfield is Dr Richard Benton, an ecologist from the University of Manchester. He points out that 'muck and magic'—the almost intuitive application of experience to gardening does work in the green fingered hands of the experienced gardener. For the rest of us doing what gardening books or personalities tell us without knowing why, can lead to disaster and is, in any case, harder to apply than doing what we know supplies the plants' needs. It's a more scientific approach—but don't be put off.

Dr Benton has some quietly revolutionary approaches too that you might like to consider—especially as they do away with much of the more tedious aspects of gardening. He is, for instance, experimenting with broadcast sowing. In competition with a traditional, professional, gardener he grew almost as many carrots by scattering the seed over the plot instead of sowing in lines. As a result the carrots' leaves covered the soil, supressed weeds and kept the root run moist. As the plants got big enough to pick he pulled them up leaving room for those around to grow bigger. His plot had four visits in the season; his rival's was hoed, thinned and weeded regularly. This year I'm giving up straight lines!

Dr Benton is also unconvinced that weeds are the enemies of crops. Few, he reckons, really compete for the same nutrients, most provide shade for the soil, some may actually extract nutrients the food plants aren't able to convert and make it available for the next crop.

A great many experiments remain to be done—some may be tried at Macclesfield—but the general pattern seems to be a move away from puritanical gardening to a more natural approach that will give results without calling for the regimented perfection that frightens some of us to death!

As far as the Grants are concerned, they are fairly certain to be self-sufficient in vegetables by the end of the first year and in most fruits before long. If food prices continue to rise the way they have been doing they may end up saving more on food than they do on fuel!

Summary

Bear in mind, that these practical suggestions are taken from one collection of experiments and the experience of one small group of people. We may have thought about, read about and worked on little else for almost two years but it's a big subject and we wouldn't like to claim that our solutions are definitive. What our experience underlines is that you should know *why* things happen. Once you understand the basic rules you'll be less likely to make expensive mistakes. Arnold Stead, chimney sweep, double-bass player and windmill enthusiast designed his most successful mill in his head every night for one hour in bed—for five years! So have a go—but while you're at it, do it well!

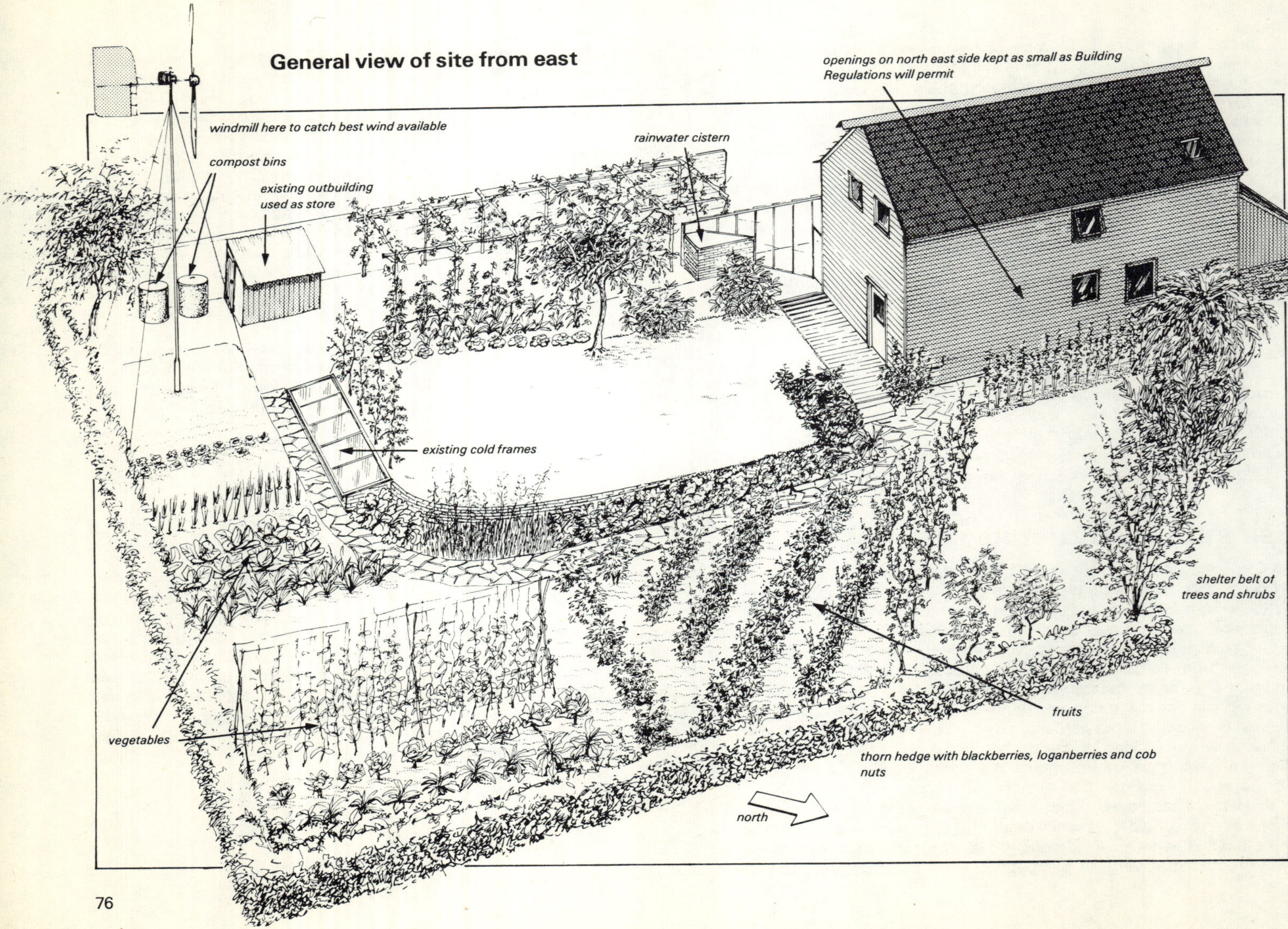
General view of site from east
openings on north east side kept as small as Building Regulations will permit
windmill here to catch best wind available
rainwater cistern
compost bins
existing outbuilding used as store
existing cold frames
shelter belt of trees and shrubs
fruits
vegetables
thorn hedge with blackberries, loganberries and cob nuts
north

Further Information

Suppliers

Solar Panels The International Solar Energy Society (UK Division), Royal Institution, 21 Albemarle Street, London W1, publish a list of manufacturers and consultants on installation.

Solar Panels, Wind Generators, etc. Conservation Tools and Technology Ltd., 143 Maple Road, Surbiton, Surrey KT6 4BH, manufacture and import a wide variety of equipment.

Solar Batteries Joseph Lucas (Sales and Services) Ltd., Birmingham B18 6AU, Ferranti Electronic Components Division, Gem Mill, Chadderton, Oldham OL9 8NP.

Methane Digesters Bio-Gas Plant Ltd., 75 Stamford House, Stamford New Road, Altrincham, Cheshire.

Books

General

Energy, Environment and Building, Philip Steadman, Cambridge University Press.

The Autonomous House, Brenda and Robert Vale, Thames and Hudson.

Survival Scrapbooks 1-5, Unicorn Books, Gloucester Road, Brighton.

Solar Power

Solar Energy for Man, B.J. Brinkworth, Compton Press.

Solar Energy and Buildings, S.V. Szokolay, Architectural Press.

Wind Power

The Generation of Electricity by Wind Power, E.W. Golding, Spon (out of print but available from libraries).

Windmills and Millwrighting, Stanley Freese, David and Charles.

Water Power

Windmills and Watermills, J. Reynolds, Praeger.

Water Turbines, Paul N. Wilson, HMSO.

Methane

Methane, Fuel of the Future, Bell, Bolter, Dunlop and Keiller, Singer Books, Cambridge.

Acknowledgments

Many thanks to Polly Bide, Brian Trueman and Don Wilson without whom this book would not have been possible.

The publishers would like to thank the following for permission to reproduce the photographs in this book:

Allan Baker, page 24; *Douglas Dickins,* page 28; *The Dow Chemical Company,* pages 55, 56, 58, 59; *Malcolm Goy/City University, London,* page 34; *Granada Television,* pages 4, 5, 7, 8, 12, 21, 45, 70, 71, 72, 73, 78; *Keystone Press Agency,* pages 16, 22; *T.P. McLaughlin,* pages 25, 29; *Dr Peter Musgrove,* page 32; *National Trust/Olive Kitson,* page 29; *Paris Match,* page 21; *Photri,* page 49; *Popperphoto,* page 42; *Royal Naval Survival Equipment School,* page 43; *Street Farm,* page 31.

Back cover: top left, by courtesy of AGA, thermovision photograph; top right, Brian Trueman examining solar panels at Pilkington Brothers Research and Development Centre; bottom left, Brian Trueman's model Cretan windmill; bottom right, solar roof being installed at Macclesfield house.

Brian Trueman and architect Donald Wilson

Index

Baer, Steve *18*
Bate, Harold *40*
Beatson, Robert *31*
Benton, Dr Richard *75*
Bio-gas *39-40*
BRAD (Biotechnic Research and Development) *9, 12, 46*
Brinkworth, B J *17*

Cartwright, Dr Bill *72, 73*
Chemical storage *44*
Clivus Multrum toilet *41*

Darrieus, G J M *31, 33, 74*
Double, triple glazing *51*
- *sealed unit 52*
- *supplementary frames 52*

Dunn, Professor P D *40*

Financing *Building Societies 48-9*
Flat-plate collector
- *definition 9*

Francis turbine *34, 35*
Fry, L John *75*

Garden, self sufficiency in *75*
- *at Macclesfield 76*

Gee, Harold *38*
Glazing *51-2*
Grant, Geoff *5, 6, 69*
Grant, Lynn *5, 6, 61*

Heat pump *36-8, 46, 49, 74*
Hills, Lawrence *75*
Hydraulic ram *35, 46*

Insulation
- *ceilings 53*
- *doors 50, 53*
- *floors 56-7*
- *need for 50*
- *roofs 53-5*
- *walls 57-61*
- *windows 50-2*

Low-grade heat, use of *36-8*

Macclesfield house
- *conservatory 19-21, 44, 45, 53*
- *heat pump 74-5*
- *insulation 54, 56, 59-61*
- *plans 64, 65, 67, 76*
- see also *solar roof*; *windmill*

Methane gas *75*
Methane generators *39-40, 48*
Mullbank toilet *41*
Musgrove, Dr Peter *31-2, 74*

NASA, windmill *33*

Odeillo valley solar energy experiment *21-2*

Pelton wheel, *34, 35*
Pike, Alexander *18, 44*
Savonius, Sigurd *31*
- *rotor 31*

Shutters *51*
- *do-it-yourself 61*

Soil Association, The *75*
Solar batteries *23-4*
Solar cooling *23*
Solar energy
- *collecting 9-14, 17-18, 62-3, 68-9*
- *storing 18-23, 44, 63-8, 74-5*
  - *rock store 19, 20*
  - *solar batteries 23-4*
  - *wall store 18, 21*
- see also *solar panels*; *solar roof*; *solar stills*

Solar panels
- *and heat pump 37*
- *circulating the water 14-17*
- *closed system 9, 13-14*
- *commercial 7, 14-15, 48, 49*
- *homemade 15, 48*
- *'open' system 9, 14*
- *planning permission 46-8*
- *siting of*
  - *ground-level 15-16*
  - *roof 15*
  - *wall 16*
- *use of 62-3*
  - see also *storage of energy*

Solar roof *48*
- *at Macclesfield 8, 10, 11, 17, 18, 19, 46, 63, 70*
- *planning permission 46-8*

Solar stills *43*
Stills, solar *43*
Storage of energy
- *problems 44*
- see also *solar energy*; *chemical storage*; *wind power*

Thermo-syphon *68-9*
Thermovision *50*
Triple glazing *51*
Turgo Impulse Turbine *35*

Vale, Robert *9, 46*

Waste, recycling *41*
Water
- *butts 42-3*
- *recycling 42-3*
- *stills, solar 43*

Water power *34-5*
- *planning permission 46-9*

Water turbine *34, 35, 46, 49*
- see also *Francis turbine*; *Pelton wheel*; *Turgo Impulse Turbine*

WESCO (Wind Energy Supply Company) *33*
Wilson, Donald *6, 9, 75*
Wind generator *25-6, 28-30, 48*
Windmills *28-31*
- *commercial 74*
- *horizontal-axis 33*
  - *canvas 25, 28, 30*
  - *Cretan*

at Macclesfield 70 - 74
multi-bladed 30
slat 28, 29
twin-bladed
Macclesfield 33
vertical-axis 30-1, 32
Darrieus 31, 33, 74
Musgrove 31-2, 74
Savonius rotor 31
Wind power 25-8
future of 32-3
planning permission 46-8
storage of energy 32, 44, 74-5
use of 70
Winn, John 72, 73